Ecological Agrarian

Ecological

Agrarian

✿ _____

Agriculture's First Evolution
in 10,000 Years

J. Bishop Grewell and Clay J. Landry
with Greg Conko

✿ _____

Purdue University Press
West Lafayette, Indiana

Library of Congress Cataloging-in-Publication Data

Grewell. J. Bishop (Justin Bishop), 1975–
    Ecological agrarian : agriculture's first evolution in 10,000 years /
J. Bishop Grewell and Clay J. Landry with Greg Conko.
        p. cm.
    Includes bibliographical references.
    ISBN 1-55753-296-6 (alk. paper)
    1. Agricultural ecology. 2. Agricultural ecology—United States. I.
Landry, Clay J., 1968– II. Conko, Gregory P. III. Title.
S589.7.G74 2003
577.5'5—dc21

                                                            2002155039II

PRINTED IN THE UNITED STATES OF AMERICA

To Terry L. Anderson

for taking a chance on two guys from
south central Montana.

Contents

Acknowledgments

Behind the scenes of any book are support staff, support families, financial supporters, and a host of other contributors. As books are inherently personal things, the line between all of these is often blurred.

We would like to extend thanks to Greg Conko for contributing his excellent chapter on biotechnology. We would like to thank Jane Shaw and Roger Meiners for providing helpful comments on earlier drafts of this manuscript. This book would not have been possible without the financial wizardry of Terry Anderson and Monica Guenther. We thank them. As always, the support of the great people who work at PERC's office in Bozeman, Montana, made the days fly by and, more than once, they spurred ideas on these pages. Four interns tracked down stories for us during production. Clay's "mini-me's" were hard workers, which we were grateful to stumble across. Thanks to Mykel Matthews, Deb Jacobs, Kris Kumlien, and Ryan Anderson.

Idea entrepreneurs must have idea venture capitalists willing to invest in them. In that vein, we wish to thank the foundations and individual supporters to PERC that made this project happen. In particular, the Earhart Foundation offered generous support that was pivotal to this project, as did two anonymous donors. Much gratitude is owed the staff of Purdue University Press for bringing this book to market, including our author liaison Donna VanLeer, director Thomas Bacher, managing editor Margaret Hunt, and the editorial board for deeming our book worthy of the Press's resources.

Without the eco-entrepreneurs leading the revolution and evolution discussed here, we would have nothing to write. Thanks to those whose stories we told and to the countless others we haven't even discovered. Special thanks to those ecological agrarians who helped us tell their stories.

Finally, Bishop would like to thank his parents, Betty and John, and his sister, Jenny, for their lifelong support. And he would like to dedicate this work to his Grandma Eunice, who kept her kids close, and his Grandpa Cliff, who deserved a ranch of his own. Clay would like to thank his parents, Clay and Jamie, for instilling him with a lifelong appreciation for agriculture and an undying interest in finding solutions rather than problems.

1.

AGRICULTURE'S ENVIRONMENTAL TRIUMPH

Agriculture is undergoing its first major evolution of purpose in 10,000 years. Shifting from the guiding light of feeding people's bodies to feeding their growing desires for a healthy planet, it is evolving into a new entity. As the quiet backbone of civilization, it brought us out of nature. Now, it undertakes the no-less daunting task of preserving the wilderness where humans were born. This book offers a glimpse at where the new future hopes to take us and how we will get there.

For ten millennia, humans struggled with agriculture to cultivate enough food to feed them. Their mornings, evenings, and afternoons were spent sowing seeds in the soil, dredging through mud, carrying water on their backs, domesticating livestock, and fighting off unwanted weeds that threatened crops. Praying to the deity of the day, they beseeched the gods for rain during growing seasons and asked for fair weather in harvest. Work was long and hard. It filled the waking hours.

Survival was the payoff.

While history was written, the quest to feed the planet moved quietly forward. Great civilizations arose, achieved glory, and settled back into the dust of their forebears. Wars over land and resources were fought. Often food played both the culprit and the prize. Languages developed and died. Through it all, farmers and ranchers toiled toward a single goal. They sought to fill their belly and that of their tribe, countrymen, and even foreigners if the price was right.

Starvation was a daily reality. The survival of a lone milk cow could nurture a family through the winter as sure as the animal's demise would lead to death's embrace. The Grim Reaper's employment of a scythe to harvest lives made the thin line between life and death clear; agriculture's success or failure determined who would live and die with more accuracy than any other factor.

Battling to feed the planet in the face of the elements outlasted every war in history. For epoch spans, it seemed the battle might never be won. Humans would continue to struggle until their final day or hunger would win out. As late as the 1970s, this view held popular sway.

In 1968, two million copies of Professor Paul Ehrlich's *The Population Bomb* sold the premise that famine was destined to win. The burgeoning world population was too much for food production to handle. He wrote, "The battle to feed humanity is over. In the 1970s, the world will undergo famines—hundreds of millions of people are going to starve to death in spite of any crash programs embarked on now."[1] Ehrlich wrote two years later that 65 million Americans and 4 billion other people would die from starvation between 1980 and 1989.[2] Others agreed with Ehrlich and an international outcry arose. We were doomed and we could only hope to minimize the damage.

But Ehrlich and his cohorts were wrong in every way but one. The battle to feed humanity is indeed over—except humanity won. Victory, however, leaves a gaping question. What next? Where does the future challenge lie? What is the new goal for agriculture? Are we done?

We are not done. But what is next? After any successful war campaign, the next job is to restore what was destroyed on the battlefield. Agriculture defeated hunger, but it harmed aspects of the natural environment in the process. Irrigation run-off polluted waterways. Cultivation of land infringed upon wildlife habitat. Pesticides damaged more than just pests.

Cleaning up the battlefield while continuing to provide food and meet new environmental demands is agriculture's evolving task. It must deal with the problems that arose in the process of feeding the world while holding the line of its past success. This book is about agriculture's new phase.

Agriculture is undergoing an evolution. In the twenty-first century, farmers and ranchers still must feed the world, but it is no longer the primary problem. The world is fed. Agriculture must now feed the world while helping the environment. The public demands it.

In a sentence, this book unearths agriculture's evolution and its entrepreneurs—eco-entrepreneurs—who are producing food *and* protecting and nourishing the environment at the same time. The well-fed populous demands environmental amenities and is willing to pay for them. As the wealthiest country in the world, the United States is where many of the changes are appearing first. Therefore our book focuses on the evolution in this country. To capture these new demands, agriculture in the U.S. (and abroad) is transforming itself. No longer is it sufficient to produce food; it is now important how that food is produced. Agriculturalists who wish to flourish must meet the growing demand for a healthy environment.

This volume asks how and why agriculture's evolution is taking place. We contend, with the problem of hunger solved, agricultural entrepreneurs are now responding to the populace's increasing demand for environmental protection. Discussing agriculture's evolution to meet these demands, we ask what effects outside influences—legal rules, government programs, and increasing centralization of agriculture—have on eco-entrepreneurs working towards agriculture's new form. By asking where agriculture is going next, we hope to help it complete the current evolution in less than the 10 millennia it took for the last evolution. But in order to see where we are going, we must first figure out where we have been.

From Where Did Our Agriculture Arise? A Brief History

While agriculture arose independently among the native population of the United States, the crops domesticated by Native Americans are not the crops raised in North America today. Native Americans began to domesticate plants around the period of 2500 to 1500 BC in the eastern U.S. The natives nurtured sunflowers, a relative of spinach called goosefoot, squash that provided both edible seeds and small containers, and sumpweed. Around 500 to 200 BC, knotweed, maygrass, and little barley were added to the mix. It wasn't until the two hundred years surrounding 1000 AD that corn and beans from Mexico joined squashes to form the trinity that would provide the majority of Native-American agricultural production. These three plants and their relative advantages over the natives' first crops pushed out the ancient crops so that by the time Europeans came to America, only sunflowers and eastern squash remained from the initial eastern U.S. garden.[3]

While corn remains a primary crop today, our other major crops were not around until after European colonization: wheat, barley,[4] and soybeans were brought from other countries. Wheat and barley, in particular, came from Europe along with most of our domesticated animals.[5] But these products of agriculture did not start in Europe, either. They found favor with the beginning of agriculture in the Fertile Crescent of southwest Asia. Around 8500 BC, in the region now occupied by Iraq, Iran, and Jordan, agriculture began.[6] The domesticated plants of the Fertile Crescent spread to Greece by 6000 BC. Over the course of the next thousand years, they were adopted by the societies covering much of mainland Europe. Finally, England, France, and the Scandinavian countries joined the sedentary lifestyle of farmers and ranchers by about 3800 BC.[7] Crops and animals were genetically modified through breeding (both intentionally and unintentionally) for qualities humans desired, which resulted in a self-feeding cycle of more crops and livestock leading to more humans who spread to further regions and planted still more crops and raised more livestock.

It was the success of agriculture that allowed for civilization. In his Pulitzer Prize winning work, Jared Diamond writes of food production, "It became possible, for the first time in human evolution, to develop economically specialized societies consisting of non-food producing specialists fed by food-producing peasants."[8] Agriculture led to civilization in three ways.

First, the sedentary lifestyle of agriculture freed people to stay in one place and accumulate possessions. In the nomadic lifestyle of hunter gatherers, continuous travel limited one's goods to the bare necessities: carry-on luggage only. "You can't be burdened with pottery and printing presses as you shift camp," points out Diamond.[9] And yet possessions beyond the essentials are the foundations for building a society of records and technological advancement.

Second, specialization in food production combined with trade allows some members of a group to work on things other than feeding the tribe. As a few members of the tribe specialize in making food, they are able to produce not only enough food for their family, but a little extra. They trade this extra food in the marketplace for other goods. In turn, other members of the tribe buy their food in the marketplace and specialize in occupations that do not involve directly producing the food they will eat. Historians, inventors, blacksmiths, and other trades of civilization begin to evolve.

Finally, there is evidence that agriculture requires a more disciplined lifestyle that demands certain advances to continue. Thus, certain technological advancements are created by necessity. "It has been suggested, for example, that writing may have come into existence because records were needed by agricultural administrators," writes Charles Heiser.[10] The majority of tablets recovered from the ancient Sumerian civilization that flourished around 3000 BC were clerical records of "good paid in, workers given rations, and agricultural products distributed."[11]

Regardless of which component of agriculture led to the diverse occupations and technological advancements known as civilization, there is little doubt that agriculture is responsible for society's ability to provide "thinkers." These inventors, policy makers, and philosophers arose as a result of others spending their days in the fields to provide food for everyone else. While thinkers are known for providing solutions, they must first identify problems.

The first problem for humans was putting food in their belly. For most of the two million years that humans have lived on the planet, hunting and gathering provided the answer. Hunting and gathering solved the first problem, but it did not allow humans to do much else beyond eat, sleep, procreate, and die. Agriculture gave humans a life beyond their biological needs. After a couple thousand years of experimentation, cultivated crops provided more caloric output per unit of time and space than hunting and gathering did.[12] This allowed a portion of the population to do something other than hunt for food. Still, solving one aspect of a problem can create other problems or simply change the first problem. Agriculture was, and is, no exception.

Agriculture, Its Advances, and the "First Environmental Apocalypse"

With the development of agriculture came the identification of the first perceived "apocalyptic" environmental problem. Increased food production due to farming and ranching allowed the human population to grow faster than it had before. With a greater population came more mouths to feed. The two were intertwined in a race. More food leads to more people who then demand more food, which leads to more people. . . . The battle with which we opened this book had begun.

One of the first great pessimists and predictors of modern environmental disaster arose from the study of the cycle between increased population and increased food production. In 1798, Thomas Malthus published

An Essay on the Principle of Population. In it, he conjectured that while food production grew as an arithmetic progression (1, 2, 3, 4, etc.), population grew geometrically (1, 2, 4, 8, etc.). Hungry mouths would always outpace meals to feed them. If this was true, Malthus predicted there must be checks on population growth in one of two ways. Either preventative control through moral constraint and marriage deferment would reduce population growth or the direct checks of famine, war, and the other horsemen of the Apocalypse would increase death rates.[13]

Malthus forgot one important variable. "He ignored the possibility of technological advance in agriculture allowing a greater increase in food output," writes David Grigg.[14] By leaving technology out of the equation, Malthus set his theory up for failure. Both the history preceding him and the history following proved him wrong. Long before the great thinker Malthus had identified the problem, agriculturalists were working to solve it. Contrary to Malthus' dreary forecasts, food output stayed ahead of population growth.

The first major advancement—after the advent of agriculture through domestication of plants and animals—arrived with irrigation. Irrigation commenced in the Fertile Crescent region around 5000 BC and spread across the Eurasian continents over time.[15] It developed separately in Mexico around 1000 BC. Great power has been attributed to irrigation. Writing of the Fertile Crescent, agricultural historian N.S.B. Gras claimed, "In Mesopotamia where the calcareous soil is fathoms deep, there was a yield of 200-fold, and probably without exhaustion or impairment as long as the water of the Euphrates and Tigris were kept in control and properly distributed."[16] Supporting these claims, K.D. White offers, "The establishment of the first sizeable concentrations of sedentary farmers took place in Mesopotamia, where flood-control was essential to economic survival; the rulers of Sumer were the pioneers of large-scale irrigation, while at a somewhat later date the Nile Valley and its Delta were the scene of elaborate developments in the field."[17] When the Sumerian, Babylonian, and Egyptian civilizations faded away, the Greeks took up the practice.

It should be no surprise that the Greeks were followed in irrigation by the ancient Romans, who became famous for their aqueducts. Rome developed flood control, dams for water storage, and of course, irrigation. Cato the Elder, writing in the middle of the second century BC, discussed the importance of draining and clearing ditches because they otherwise might flood and damage grain fields.[18]

Irrigation's importance in keeping food production ahead of hungry mouths is no less important today. Indur Goklany, of the office of Policy Analysis at the U.S. Department of the Interior, credits irrigation with tripling productivity on average.[19] Irrigation played and plays an important role in agriculture's response to the first great environmental problem.

Fertilizers, too, played an important role in the battle to outpace famine. Colonial Americans increased food supplies by employing oyster shells, lime, gypsum, and organic waste in fertilizing their fields.[20] Fertilizer has been around nearly as long as farming.

> Perhaps most reliance has been and will be placed upon fertilizers, both commercial and barn-yard. In every stage of agriculture we find the use of fertilizers though some peoples and individuals may have neglected them. The Indians of New England used fish heads to plant along with their corn in natural husbandry, and the English used marl as an ameliorator and barn-yard manure as a fertilizer in the Middle Ages when practising the naked-fallow system. And, of course, in succeeding stages, more fertilizers and manures were applied than before, especially at the hands of able cultivators.[21]

The use of fertilizers in America was influenced by the colonists' European background. Europeans well understood the advantages of fertilizer and the immigrants to America brought that knowledge with them.[22] As with most of their agricultural knowledge, the Europeans inherited some of their knowledge from the ancient Romans who no doubt learned from earlier farmers in Sumer, Egypt, and Mesopotamia. "The main resource of the Roman farmer in replenishing fertility was farmyard or stable manure," claims White.[23] Columella, writing in the first century AD, emphasized the importance of fertilizer to the Roman farmer. He offered composting as a substitute for animal manure, "in places where neither cattle nor fowl can be kept. Even in a place like this, however, it is a sign of a lazy farmer to be short of fertilizing material."[24]

Today, manure is still a major source of fertilizer, but synthetic fertilizers find increased use, too. With man-made chemicals, humans can synthesize nitrogen and phosphorus for increased production. Nitrogen application in the U.S. increased by a factor of 18 between 1940 and 1970, rising from 419 million tons to 7,459 million tons. Potash went from 435 million tons to 4,035 million tons.[25] Fertilizer has doubled yields in some cases.[26]

Mechanization helped win the battle, too. Ploughs tilling fields ensured better growth of seeds into crops. Improvements in agricultural implements increased what agriculturalists could accomplish in a day's work, leading to better yields. Author Peter McClelland explains how better equipment for plowing, harrowing, cultivating, reaping, and winnowing led the fledgling United States down the road to international power.[27] He argues these implements were the most important advancement in the American agrarian revolution. "No other agricultural feature has undergone such radical improvement between colonial times and the present as has occurred in the case of farm machinery," claimed another observer writing even before major improvements in tractors, harvesters, and other heavy machinery occurred.[28]

The variety of farm implements and the number of farm machines has been escalating for some time. Farm machine numbers rose from 300,000 in the U.S. around 1940 to 3 million in only twenty years time.[29] Technological improvements drove America's machinery boom. As the machines did more for farmers, the farmers bought more of them.

> The tractor was greatly improved in the 1930s. Its weight relative to its horsepower was reduced. Iron or steel wheels were replaced by pneumatic rubber tires, which not only made tractors more comfortable to ride but also reduced fuel consumption significantly. Gear ratios were changed, giving tractors a road speed of up to 20 miles an hour and a field speed of between 3 and 5 miles per hour. As a result of these improvements, the number of tractors on farms increased greatly. In 1940 there were 1.6 million tractors in use on farms; this number increased to 3.4 million tractors in 1950 and to 4.7 million in 1960. The work horse had virtually disappeared on farms by 1950.[30]

Inventors were creating better machines long before these modern improvements. The watermill and windmill gained adoption for grinding wheat during Europe's medieval period. Improved yokes for plough animals, the replacement of oxen with horses, and increased use of iron also moved forward at this time.[31] Before the Dark Ages, Romans employed ploughs attached to a beam yoked between two oxen to open the ground. Simple hoes cultivated the fields and kept weeds out of Roman fields. Sickles harvested grain. Winnowing baskets and shovels separated the wheat from the chaff.[32]

Mechanization in the modern age reduced the need for animal and human labor, as tractors guided by one person replaced numerous horse-and-mule drawn ploughs guided by several people. This reduced the cropland needed to feed animals such as horses and mules. It also decreased the incentive for high birth rates, by lowering the demands for human labor.[33]

Pest control systems, too, have been around for a long time in the battle to defeat starvation. Fallowing fields, which was probably developed in the Fertile Crescent, was the first major system for fighting back pests.[34] By eliminating the crops that the bugs relied upon in one area, the pests would die out instead of building their numbers year after year. After that first improvement in pest control, it took centuries before new advancements came along.

Today, our development of chemicals into pesticides greatly reduces the loss of crops to bugs, weeds, and fungi. Oerke et al. found that 42 percent of the world's crops are lost each year to pests, but that number would be 70 percent without pest control.[35]

In addition to irrigation, fertilizers, mechanization, and pest control, improved seeds led agriculture's march forward. The continued cross-breeding and genetic manipulation of plants and animals led to yields improving year after year. The progression of seeds was a pivotal stepping stone on the road to civilization.[36]

Harvesting Progress

Higher yields from the technological innovations above resulted in increased food production with less input from human labor. This reduced the work needed for humans to obtain food. Much of this advancement occurred in the last 200 years. At the beginning of the nineteenth century, a family of five in cosmopolitan Berlin, Germany spent 72.7 percent of its budget on food.[37] Nearly two centuries later in the United States, only 14.2 percent of total personal consumption was spent on food.[38] Our food today costs a third of what it did just four and a half decades ago in 1957.[39] Sustenance has become cheap in terms of the family budget. Cheap food indicates agriculture is winning the race to feed the world.

Agricultural advancement can also be put in terms of hourly human labor. Grain represented half of the food basket around 1800.[40] At the time, the average French worker spent 1,200 hours to provide for his family of four's wheat needs.[41] With the peasant or working person eating between two and three pounds of bread per day,[42] a family of four ate about 10 pounds each day or 3,650 pounds per year. Thus, a pound of bread cost roughly 33 minutes of labor.[43]

Figure 1.1. Minutes Worked to Purchase One Pound of Bread

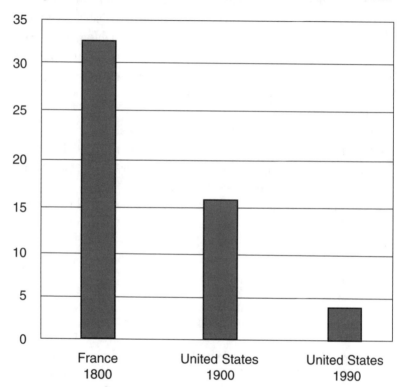

A little over a century later, the work time required to purchase a pound of bread in the U.S. had been halved from that of nineteenth-century France to 16 minutes. In 1990, the work time had been quartered again to four minutes.[44] Wheat prices have fallen ever since 1800, making it ten times cheaper on average to buy the golden grain now than it has been over the last 500 years.[45] Never in the history of the world has the average human been as wealthy in terms of the items bought for their labor as he or she is today.

During the last half of the twentieth century alone, there was a rapid decline in the human labor component of farming. In the U.S., human farm labor declined by 26 percent during the 1940s. This could be attributed to the war effort, except that it fell 35 percent in the 1950s and another 39 percent during the 1960s.[46] The likely culprit for these productivity gains that fed more of us with less labor was the internal combustion engine.

Victory!

What then was the result of this long-fought war to feed the world, to keep food production ahead of population growth? At the start of the twenty-first century, the battle appears won. Worldwide population is leveling off. Humans have become wealthy (and thus healthy) to the point where they don't need to have lots of children to ensure their survival.[47] They have improved technology to the point where they don't need as much human labor to produce their daily bread.

It is now estimated that the world population will peak at 8 to 8.5 billion around the middle of this century and then steadily decline.[48] The low range of the United Nation's predictions has population dropping off before it reaches 8 billion, while the mid-range estimate predicts population will pass 9.3 billion in 2050 before leveling off at 11 billion in the year 2100.[49] More important, we are feeding the people of the world and have the technological ability to feed the extra mouths scheduled to arrive. Jesse Ausubel (2000) of Rockefeller University estimates that if the world's average farmer reaches the yield of the average, not the best, U.S. corn grower, it will take only half of today's current cropland to feed 10 billion people. The Food and Agricultural Organization of the United Nations (FAO) and numerous other governmental food agencies predict further increases in yields above population growth for those coming decades as well:

> . . . the FAO predicts that there will be more food for more people, both in 2010, 2015 and 2030. It is expected that there will be fewer malnourished people, and that all regions will experience increasing available calories per capita…The same general conclusion is reached by IFPRI, USDA and the World Bank—and all three predict ever lower prices.[50]

According to Nobel-winning economist Amartya Sen, no famine has ever taken place in a functioning democracy whether rich or poor.[51] Dennis Avery argues that today's incidences of hunger occur only in areas suffering from civil and political strife.[52] Regardless of where hunger occurs today, it does not stem from a failure in production of food, but rather from distribution of food. There is enough meat and potatoes, rice and corn, cassava and bananas to feed the world, it is only a matter of getting that food to hungry individuals.

Agriculture has solved its first problem. No longer must people starve for lack of food. No longer must they toil in the fields day and night to put together enough staples for survival. No longer is one bad harvest the difference between life and death. Agriculture has won its race with population. Today, it produces enough food for every man, woman, and child. While 10,000 years to solve the problem seems long, progress has been rapid. After all, it took nearly two million years to develop agriculture in the first place.

Agriculture's Current Evolution

With the food problem solved, however, new demands arose. Along the way, "The change from pure savagery to crop culture and to the care and rearing of animals was marked by another characteristic of civilized man—the acquiring of wealth."[53] Increased wealth has created demands for new items, including environmental amenities. These new demands lead to conflicts over resources. Environmental problems other than feeding the world have been identified.

Only in the last century, the twentieth century, did the concept of environmentalism take hold. For most of history, humans sought to control a hostile environment, not preserve it. The word environmentalism was not even used much before Rachel Carson's 1962 book, *Silent Spring*. Discussions about modern-day environmental topics were couched under the heading of resource conservation.[54]

While the great innovations in agriculture helped solve the first great environmental problem of sustaining human life, they followed the pattern of history, and have created a host of the new problems that lie at the heart of modern environmental concerns and agriculture's current evolution. Irrigation diverts water from rivers. It can drop flows to a level where fish and other aquatic species struggle to survive. Dams, which provide irrigation districts and routes for barge traffic carrying crops, cut off salmon from their spawning grounds. Fertilizers are the primary source of water pollution as nitrogen and phosphorus lead to eutrophication (See chapters Three and Four). *The 1998 National Water Quality Inventory* blamed approximately 60 percent of pollution in the nation's rivers and streams and 45 percent of the pollution in our lakes on agricultural sources.[55] Pesticides were the subject of Rachel Carson's book as worries grew that DDT harmed bird populations. Perhaps the biggest concern about the environment and agriculture has been land conversion. Ignored in the discussion above on improving technology to increase food yields was the most basic method used to increase food production. For most of history, man increased output by adding more land to the equation. Clearing forests, draining swamps, and converting the land available to farm or ranch was the default for gaining more food.

While in the past these things went unnoticed or were ignored in the battle to feed ourselves, we now have the wealth and time to pay attention to the consequences of these actions. Environmentalism is now an important issue and a common concern. As income rose, citizens gained the leisure time and ability to make the environment a priority. The health of water, air, wildlife, and other ecological considerations matter today. Economist Don Coursey found that above a certain threshold of wealth, a ten percent income rise leads to a 25 percent increase in the demand for environmental amenities.[56] Agriculture has begun to respond to this changed focus, offering solutions to both the new demands and the problems that arose from the battle to feed humanity. This is its evolution—its change to a new being.

Take one of the most visible environmental causes: protecting wildlife. Preserving biodiversity is an important cause for environmentalists, biologists, and, opinion polls indicate, most citizens. This can be seen in the proliferation of environmental groups with wildlife monikers: Defenders of Wildlife, the National Wildlife Federation, National Audubon Society, World Wildlife Fund, and Trout Unlimited.

Biologists agree the key to preserving wildlife and maintaining biodiversity is habitat. In this area, agriculture's search for higher yields has proven helpful to both food production and saving wildlife. By increasing yields, we feed more people on the same amount or even less land. This is crucial to saving wildlife habitat. Dennis Avery of the Hudson Institute makes it clear that higher-yield agriculture protects wildlands. Avery argues that, "Producing food supplies with 1950 technology would cost us 15 million square miles of wildlands today. Strictly organic food use would eliminate practically all of our wildlands."[57] Goklany used data from the Food and Agriculture Organization (FAO) to find:

> Had technology—and therefore yields—been frozen at 1961 levels, then producing as much food as was actually produced in 1998 would have re-

quired more than a doubling of land devoted to agriculture. Such land would have increased from 12.2 billion acres to at least 26.3 billion acres, that is, from 38 percent to 82 percent of global land area. . . . Cropland alone would have had to more than double, from 3.7 to 7.9 billion acres. An additional area the size of South America minus Chile would have to be plowed under.[58]

Nobel Laureate Norman Borlaug won the 1970 Nobel Peace Prize for his efforts to increase grain yields. He writes that on 1961 cereal yields alone, an additional 850 million hectares (2.1 billion acres) of land of the same quality would need cultivating to equal the cereal harvest of 1999.[59] Fortunately, yields in all areas—including cereals—did increase. From 1935 to 1970, wheat yields increased from 12.3 bushels per acre to 31.8. Corn jumped from 20.5 bushels per acre to 80.8. Potato production leapt from 66.3 hundredweight per acre to 226.3. Cotton went from 185.4 pounds per acre to 436.7.[60] These rises were due to the biological and chemical technological advances now known as the "Green Revolution."

Producing more food on the same amount or less land helps wild lands. By 2050, pesticides and fertilizers could save 30 million square miles of forests, prairies, and other prime wildlife habitat from conversion to agricultural use.[61] Without pesticide use for fruits and vegetables, it would take an additional 2.5 million acres to produce the same output. This equates to 44 percent more than the existing acreage for these crops.[62] The use of technology to save land is also evident in the developing world.

> From 1961 to 1966, Indian farmers required 13 million hectares to achieve wheat production levels of 0.83 tons per hectare. With the simultaneous adoption of genetically improved varieties and intensive crop management practices (including pest control and fertilization), production increased fivefold. Since the mid-1970s, the amount of land devoted to wheat production has leveled off at just more than 20 million hectares. In the absence of these innovations, by 1991 an additional 42 million hectares would have been needed to achieve the same production level.[63]

During the twentieth century, much of the world's wild spaces and other lands were inadvertently protected by the "Green Revolution." By developing superior seeds and adding to them synthetic fertilizers and pesticides along with improved irrigation, higher yields were developed.

These yields produced more food on less land. Work on hybrid corns at the beginning of the century helped to double and triple yields, leading to increased returns of 300 percent and more for early adopters.[64] In the U.S., crop output per acre (including fruits, nuts, vegetable, and hay) increased by 279 percent between 1948 and 1994.[65]

Biotechnology may be offering up a new "Green Revolution" to keep producing more crops on less land (see chapter six). In addition to decreasing land use, biotechnology is increasing the supply of fresh water available by working to create salt-tolerant varieties of plants that could be irrigated with seawater or other marginal water.[66] This will leave a healthy supply of drinking water for the rest of the planet, humans and wildlife included.

On the other side of the coin, there are those who worry that increased yields lead to a declining demand for cropland, and that means declining open space. But according to agricultural economist Bruce Beattie, "net cropland loss to urbanization has been minimal."[67] Loss of cropland in some places has been made up with increases in acreage elsewhere. This may be the result of better efficiency in land use as lands that are of marginal agricultural benefit are traded for higher producing lands.

Either way, agriculture has been doing a good balancing act of feeding more people on the same amount of land. This has prevented the unnecessary cultivation of further wild lands for agriculture. At the same time, those who want open space to be preserved can be happy that much of that cropland has not gone out of production and into urban uses. Three out of every four states have less than ten percent of their land developed. Even though one-fifth of the United States' land lies within metropolitan areas, a substantial part remains either undeveloped or in agricultural use.[68] In much of the U.S., people live in clustered cities rather than in the rural landscape.

The U.S. is also undergoing afforestation thanks to increased agricultural yields. We are adding to our forests each year. Much of the forest growth in this country has arisen because land that was cleared for agricultural uses when the country was first colonized has now gone back to a natural state. Thus, when cropland is converted to another use, it is not always urban use. According to a 2000 U.S. Forest Service report, the estimated forest land area in 1997 reached 748 million acres. This was a slight increase over the five years since 1992. The report announced, "Losses of forest land to development and other land uses have been offset by afforestation and natural reversion of abandoned crop and pasture land to forest land."[69]

Still, these increased yields are not the only environmental success. After all, as Rachel Carson's work was about the harms of pesticides, some may question the benefits of more habitat if we must suffer toxins for it. But here, too, agriculture is taking leaps forward to improve the quality of the environment while still providing the food production of the past. By increasing productivity of some inputs relative to others, the harmful (and costly) inputs can be reduced. Environmental Protection Agency statistics show the volume of pesticide use in agriculture fell by 14 percent from 1979 to 1995.[70] Dividing the different pesticides up by type, herbicides declined by 6 percent, insecticides fell by 52 percent, and fungicide use dropped by 14 percent. Other chemical pesticides dropped by four percent. Agricultural economist B. Delworth Gardner writes, "Because it is insecticides that are believed to provide the largest cancer threat, the sharp reduction in use could be viewed as significant."[71]

In addition, biotechnology offers hope for further reductions in pesticide use as plants are made more resistant to weeds and insects. The USDA's Economic Research Service found that between 1997 and 1998, genetically modified crops led to an 8.2 million-pound decrease in pesticide use. Those who grew genetically modified crops applied 17 million fewer acre-treatments of pesticides than those who did not grow genetically modified crops.[72]

Another major concern for both agriculture and the environment is loss of topsoil through erosion. This can lead to sediment buildup in waterways and reduce the productivity of land. Methods such as conservation tilling can reduce soil runoff by up to 90 percent. Farmers benefit from conservation tillage because it saves them money by lowering labor, time, fuel, and machinery costs as it involves fewer trips through the field.[73] Conservation tillage consists of keeping at least a 30 percent residue cover on the soil after it is planted.[74] This slows runoff and soil loss, while leaving the microbes and earthworms below undisturbed.[75] In addition, it cuts down the fuel requirements for tractors by millions of gallons.[76] Perhaps that's why in 1999, 186.7 million acres of the 294.7 million acres under cultivation in the United States used some form of reduced or conservation tilling method.[77] The disadvantage of conservation tillage is that it requires increased herbicide application, because weeds are more likely to survive.

A Brief Glimpse at the Future
of Agriculture and the Environment

Further declines in the use of pesticides, fertilizers, and other inputs are no doubt on the way. Improvements will not only make agriculture more efficient and productive, but also help the environment. Advancements in the use of computers and global positioning satellite (GPS) systems are being used to measure how much seed, fertilizer, or pesticide are placed on a spot of land and how much in turn is harvested. This "precision farming" increases yields while decreasing inputs. In addition, using cheap equipment such as rain gages and barometers connected to computers helps farmers know when to apply pesticides and what kind to apply so as to minimize environmental harm and maximize output.[78]

For instance, Rick Hartley runs a spraying business near Attleborough, Norfolk in the United Kingdom. Before crops are up, Hartley identifies patches of thistles that can be seen when he drives by with his sprayer. When he reaches a patch of thistles, he douses it with herbicide and then turns his sprayer off until arriving at a new patch. Applying the herbicide only where needed lowers expenditure of the chemical. When interviewed by a U.K. farm weekly, Hartley explained, "While it is easy enough to do that the first time through the crop, the second dose of the split treatment goes on when the beets are up and the thistles largely hidden. As a result, you can not be as selective without the risk of misses."[79]

Before precision farming, Hartley's second trip through the beets required him to spray everywhere to make sure he got all of the thistles. Today, Hartley marks the locale of thistles his first time through with a GPS system. When he returns for the second spraying, the GPS directs where and when the sprayer should be turned on and where and when it should be shut off. This saves on herbicide and lowers cost. Such patch spraying has the potential to lower herbicide costs by up to 50 percent.[80]

Other precision farming techniques employ infrared systems to monitor yields. By scanning a field, infrared provides information on plant density. Computers using geospatial information systems (GIS) turn this data into digital maps. In places where growth is low, farmers can change their mix of inputs to try for a better result. Linking soil samples taken in a field with GPS mapping can identify where farmers and their equipment should apply more or less fertilizer. This leads to more efficient and cost-effective fertilizer application. In his book *Agriculture and Modern Technology: A Defense,* Thomas DeGregori observes, "Farmers generally do not wish to use inputs, such as seeds, fertilizer, and pesticides needlessly, since they cost money."[81] By using information as a substitute input, the other inputs can be lowered. DeGregori goes on to note that precision farming technologies are currently only practical for use by large farms in the developed world. The challenge is to make them more affordable for poorer farmers in less developed areas.

Being dynamic and responsive, agriculture is not only changing in the demands that it is meeting and the problems that it is solving, it is undergoing change in its structure. In the period from 1940 to 1998, the number of farms in the U.S. dropped from over 6 million farms to just over 2 million farms.[82] From 1970 to 1990, the number of farms dropped 27 percent.[83]

Meanwhile, average farm size rose from 174 acres in 1940 to 435 acres in 1998.[84] Whereas 92.6 percent of the farms in 1970 were taking in less than $40,000 per year in cash receipts, by 1990 that number had dropped to 70.7 percent. Farms are consolidating into larger, bigger income operations.[85]

Contrary to popular thought, this emphasis shift to larger, more intensive farms is nothing new in agriculture. In 1925, N.S.B. Gras wrote of agriculture:

> The agricultural contest may be expressed in many different ways with varying shades of emphasis. It is generally the small farm against the large farm, the homestead against the plantation, and the new and improved system against the old. It is the new system that wins. And this new system generally grows up in old lands competing with the old system. The victor is almost invariably a more intensive form of agriculture.[86]

Compare those words to Samuel Staley writing at the end of the twentieth century:

> . . . agriculture is changing. Farms are getting larger, and small family farms are increasingly less financially viable. Most family farms, in fact are supported by income from off the farm. Nationally, 49.7 percent of farmers report an off-farm source as their primary income, and 37.1 percent said they worked 200 or more days off the farm. Slightly more than 25 percent of the nation's 1.9 million farms earn more than $50,000 per year in sales, while the average net-cash return is just $22,260 per farm. As

land becomes less and less important as a source of productivity, farms are becoming even more vulnerable to competitive pressures to expand their operations to capture larger economies of scale or to shift uses into nonagricultural activities such as housing, offices, recreation or open space.[87]

Agriculture's Environmental Evolution

As we set out at the beginning, agriculture is evolving for the first time in its history. Its purpose is no longer only to feed people. This book focuses on that evolution toward meeting new demands from consumers. It asks how the agrarian sector is dealing with the new problems created by the solutions of yesterday and what frameworks can help it succeed.

The answers to agriculture's next stage are already out there. In part, it relies upon the old path of continuing to increase yields through technology. If it isn't broke, don't fix it. Biotechnology and precision farming are but two of the answers being researched for the purpose of increased yields. But there is also the shift toward meeting other demands beyond the simple scope of a full belly. This part of the answer is where agriculture's evolution lies.

Supplying the environmental amenities for a new generation of demands offers a chance for agriculture to expand its revenue. For many farmers and ranchers, the only way of staying in business is turning toward green options. These options rely on a new revenue source, more efficient use of available resources, or capturing a premium price for their current products by employing production methods that have a philosophical appeal to consumers. It is not ecological leanings on the part of agricultural producers turning them green, but rather the financial rewards for doing so.

In the next chapter, "The Color of Money," we explore two of the methods mentioned. By using markets for green products, producers improve their business through new revenue opportunities. They also achieve premium prices for products certified green in their production. The chapter discusses the efforts of ecological agrarians to gain new revenue opportunities from producing green products and how premium prices and market share can be captured through green brand names and labeling.

In chapter three, "Waste Not, Want Not," we write about more efficient resource use. Farmers and ranchers take in the waste of other industries and one another as they produce their own products. Applying the same idea in reverse, industries take in waste leaving the agricultural sector.

This lowers the costs of disposal for agriculture, cleans the environment, and in some cases, the agricultural producer often receives payment for the waste materials. Marketplace recycling turns one business's trash into another business's treasure.

Chapter four unearths the efforts of landowners and land trusts to manage the environment for the environment's sake. Unlike the other chapters, some of the landowners in this chapter are less motivated by profit and more influenced by their own sense of environmental value. Some operations in the chapter do, however, capitalize on a growing demand for agricultural tourism and fee recreation to cover their costs and generate revenue.

Chapter five turns to the habitat that agriculture provides for the nation's wildlife. In return for habitat provision, many landowners are finding revenue opportunities by providing prime hunting for a fee. In addition, environmental organizations are willing to pay landowners to leave some of their irrigation water in stream to help struggling fish populations. The chapter looks at how such new opportunities are arising and how institutional changes at the state level are helping them flourish.

In chapter six, Greg Conko documents the advances made via biotechnology to improve resource use by the agricultural sector. From plants that can live on saltwater to bug-resistant varieties that reduce the need for pesticides, biotech is turning agriculture on its toes and benefiting the environment. At the same time, it is creating a second "Green Revolution" to feed growing numbers of people without increasing the need for more land.

After demonstrating agriculture's evolution in each chapter, the book concludes with a look at where we go from here in the final two chapters. Chapter seven lists environmental harms that have arisen in agriculture's fight to feed the planet and explores institutions that have exacerbated such harms. Chapter eight synthesizes the lessons learned throughout the book to ask what institutions entice agriculture into evolving for the sake of improved environmental quality and what institutions fail. It provides a framework for nurturing agriculture forward.

Agriculture's greatest triumph in the environmental arena was feeding the world with increasing yields that protected much of our wildlife habitat from agricultural conversion. But that success was in many ways an unintentional side effect of trying to feed the world. It is now embarking on protecting the environment as an explicit goal demanded by its customers. For the first time in 10,000 years, it has a purpose beyond food production. The fresh challenge lies in maintaining the success of the past while working toward achieving new environmental endeavors. Capturing the dynamic ingenuity of agrarian entrepreneurs, agriculture's evolution says we can have our cake and eat it, too.

The stories of those carrying agriculture into its new evolution are the ones we capture in this book. We hope the eco-entrepreneurs discussed here will inspire others to plant the environmental seed through imitation or original ideas of their own. We also hope this work encourages the development of the institutions necessary for agriculture's evolution to be successful. By encouraging experimentation and rewarding those who do well, agriculture can enter into the next millennium as an important part of civilization, just as it was over the last ten thousand years. It can be more dynamic than ever. And it can do so, knowing that not only is it feeding the world's population, but it is also nurturing the planet.

The stories provided here are but a glimpse of agriculture's future. In time, their solutions may prove obsolete as the environmental problems of today become problems of the past. No doubt when they are solved, a new set of dilemmas will arise, and agriculture will evolve again. We will leave finding the answers to that next stage for the thinkers of tomorrow.

2.

THE COLOR OF MONEY

A battle is brewing in the Pacific Northwest. A fight in 2001 for water in Oregon's Klamath Basin pitted farmers and ranchers against endangered salmon and suckerfish. It is not just a play of words to call the Klamath Basin dispute a watershed event.

With water levels low from drought, the federal government cut off Oregon farmers from much needed irrigation for crops in order to maintain minimum flows necessary for preservation of the Coho salmon and two species of suckerfish. All of these fish are on the Endangered Species list.[1] In response, farmers cracked open the headgates keeping the water from their fields. When the local sheriff refused to step in, federal authorities arrived to protect the headgates from further attacks. In a short-term compromise to the farmers' revolt, Interior Secretary Gale Norton released a small additional flow of water. But as the drought continued on, so did the farmers' protest as their crops died and their losses mounted. The story was front page news across America. It challenged people to pick a side. With no long-term solution reached, the divisions created at Klamath will be more than a flash in the pan unless major changes are made.

While the water war grabbed headlines, a quiet story of mutual cooperation and success continued in the nearby Willamette Valley watershed. Developed over the years, it doesn't involve shovel brigades of farmers making late night raids for water. There are no environmentalists chanting mantras of *Agriculture must go, save the Coho!* Gun-toting government officials have not been called in to make a final decision regarding right and wrong. Yet this quiet revolution could change the dynamic of the salmon debate for years to come.

Salmon-Safe

The Hall family is a sixth-generation Oregon family, and for over eighty years, it has owned a 202-acre spread called Willamette Farms on the west side of the Willamette River. (Spring Brook Creek—a tributary of the Willamette River—runs right through Willamette Farms property.) While raising apples and hazelnuts for their livelihood, the family has always been concerned about protecting the environment around the farm. In the last few years, the Halls have found a way to merge the goals of profits and environmental stewardship into one.

Willamette Farms is part of the Salmon-Safe labeling program started in 1995 by the Pacific Rivers Council. One of the largest and most successful river conservation organizations in the United States, the Council strives to protect and restore rivers, their watersheds, and native aquatic species. In the 1980s, river guides and rafters formed the Pacific Rivers Council because they were interested in protecting the forested watersheds of the Pacific Northwest.[2] After helping to pass the largest federal water protection bill in U.S. history in 1988, the group began to look to the private sector for environmental protection.[3]

"In the Willamette Valley, non-point sources of agricultural pollution were threatening salmon," says Dan Kent, managing director for Salmon-Safe. "We realized that the advocacy tools that may have worked on public lands, litigation and public pressure, could not work as well on private lands so we took to market-based efforts. That led to Salmon-Safe."[4] In the year 2000, Salmon-Safe separated from the Pacific Rivers Council and became an independent, nonprofit conservation organization.

The Salmon-Safe label began as a way to promote agricultural products grown with practices, which protect healthy watersheds. Certification standards were developed over two years with the help of fish and aquatic scientists as well as input from agricultural interests. By certifying products that meet the group's standards, the Council hoped to help salmon-friendly farmers and ranchers capture a premium for their products. In return, it would guarantee that the enviro-sensitive production methods of agrarians such as Willamette Farms protected salmon. It worked. Growers came to feel a strong attachment to the program. When lack of funding forced the Pacific Rivers Council to end the program for a brief period in June of 1999, the farmers worked to bring it back.[5]

Each agrarian enrolled in Salmon-Safe pays a fee that covers the cost of certification and administering the program. Fees vary based on the size of the spread and the cost of the certification visit to the property. Most are between $400 and $500.[6] The enrollee's land is assessed by independent inspectors who give points to the farm based on certain criteria that lead to healthy salmon populations. For instance, a rancher who has installed fences that limit livestock access to waterways can receive positive points for protecting the waterway from cattle-caused erosion. Points are also rewarded for placing fish screens anywhere that water is being diverted from a stream, thereby lowering fish mortality. Use of the most efficient irrigation delivery systems such as diversion piping and drip irrigation instead of open ditches also scores well for an enrollee. Practices that are not conducive to fish health lose points in the certification process. To be certified, the farmer or rancher must have an overall positive score in each of five different program elements. Hence, they can be deficient in certain respects and get negative points, but still get certified because overall performance is positive.[7]

Once certified, the farmer or rancher may use the Salmon-Safe label to sell their products. The label helps the agrarian receive a higher price for their product, reach new markets, and receive positive press attention. All of these elements help improve the chance of sale and increase income for the Salmon-Safe enrollee. At the same time, endangered and non-endangered fish benefit from the new agricultural management practices.

Only family farms and vineyards are currently enrolled in the program. The size of the operations range from 15 acres to 7,000 acres. Working in Washington, Oregon, Idaho, and California, Salmon-Safe counts a total of 50 growers and 30,000 acres among its certified members. Two-thirds of the growers are vineyards, but those wineries account for less than one-third of the enrolled acreage. In addition to growers of wine-grapes, Salmon-Safe has certified five dairies (with another two on the way), several vegetable growers, the hazelnut and apple operations of Willamette farms, and a wheat and barley grower in Idaho who plans to sell his crop to beer brewers.[8]

Salmon-Safe's positive story, where everyone wins, is not the exception to the rule. It is part of a growing phenomenon throughout agriculture where the color of money and the environment are becoming one and the same. At a Salmon-Safe event in 1998, Governor John Kitzhaber identified the program as "the best example yet of the cooperative voluntary approach that is needed to prevent the extinction of wild salmon."[9]

Green Markets

Pat Dudley and her husband, Ted Casteel, run a winery in Salem, Oregon. They employ—and benefit from—Salmon-Safe's standards for vineyards. The Dudleys recognize that many consumers are looking for ways to support environmental concerns with their pocketbook. According to Ms. Dudley, buying products that protect the environment is a movement on the rise.[10]

Some consumers are willing to do the necessary research, travel to certain stores, and pay higher prices to accommodate their environmental preferences and ideologies. In a 1998 Michigan State University survey, 52 percent of apple consumers agreed that they would pay 20¢ more per pound for eco-labeled apples over regular apples.[11] Other consumers might be less willing to put in substantial effort, but will choose a product with an environmental reputation over a competitor that lacks the green distinction. In the same survey, 72 percent of apple buyers claimed they would choose eco-labeled apples instead of regular apples if both were the same price. This green demand provides an opportunity for farmers and ranchers to make an extra buck by engaging in management practices considered ecologically sound by consumers.

Often known as "green-marketing," farmers and ranchers cater to the green niche by helping fish, reducing water usage, protecting predators, and practicing other methods that appeal to green consumers. The consumers, in turn, pay a higher price or at least pick their products based on whether a producer helps the environment or not. Capturing this niche can do wonders for an agrarian's pocketbook.

Green Markets and Predator Friendly Wool

Becky Weed raises sheep in Belgrade, Montana. She and her husband, Dave Tyler, entered the sheep-herding business full time in 1993. The wool sheared and sold from their sheep often sells at more than twice the price of conventional wool. While Becky Weed's sheep produce a quality product, it isn't necessarily twice as good as traditional wool for producing hats, coats, socks, or slippers. Consumers pay more for Becky Weed's wool because they like the way she raises it.

Most shepherds deal with predators by shooting, poisoning, or trapping. Roaming coyotes and mountain lions are considered enemies destroying sheep and the agricultural livelihood. Becky Weed takes a different view. She welcomes the predators as an important part of the natural environment. As *Time* magazine wrote, Weed "has a genuine concern for the coyotes, mountain lions and bears that roam the Big Sky Country, and she believes they are as important as other wildlife for keeping the environment in balance."[12]

To forward her view, Weed joined a cooperative of shepherds in Montana and Idaho who sell Predator Friendly Wool. She and her husband reject lethal means as a way of dealing with predators. Instead, they rely on a llama named Cyrus.[13] Sheep tend to think of llamas as great big sheep. Thus, they accept the big South American animals, who protect them from predator attacks, as part of the herd. A llama is no guarantee that a predator won't kill a sheep, but it does lower the risk. As Weed notes:

> On our own place, our guard llama has been extremely successful at keeping the coyotes at bay, but we lost four ewes to a bear who came into a back field several nights a few years ago. This year we have lost several ewes and lambs to both bears and coyotes. Last February, a bald eagle flew into a pasture of pregnant ewes right behind our house and picked up newborn lambs. He ate them in an adjacent pasture.[14]

Losses that Predator Friendly producers suffer from predation are made up for in the higher prices received for their product. Wool from the co-op often sells at $2 a pound when conventional wool prices fluctuate anywhere from 40¢ per pound to $1.50 per pound.[15] Predator Friendly Wool generally sells as $130 sweaters and $165 blankets[16] everywhere from boutique stores in California, Minnesota, and Washington to online sales reaching Korea, England, and Khazakstan.[17]

During 2001, Becky Weed and her husband have met the wool needs for their line of Predator Friendly products with a herd of 210 ewes. In the past, they have bought some wool from members of their cooperative to supplement their wool supply. The couple hopes to grow their herd to no more than 450–500 animals, which they will feed with their own 160-acre spread and by leasing additional hay and pasture land.[18]

Members of the cooperative have received plenty of free publicity for their products. Stories on Predator Friendly Wool have appeared in the *Christian Science Monitor, New York Times,* and *High Country News.* National Public Radio's *Living on Earth* ran a segment on the concept. The most recent coups for the group came when Weed was featured as one of *Time* magazine's "Wildlife Heroes of the Planet" in February of 2000.

In Montana and other western states, the battle between sheepherders and coyotes is similar to the one between fish and irrigators in Oregon, except that it has been going on longer. The opposing sides are more firmly entrenched. Because of that, there are many sheep herders who adamantly oppose the idea of predator-friendly wool. Bob Gilbert of the Montana Wool Grower's Association claims Predator Friendly is "nothing more than a scam to lull the public into believing that letting predators run free works." He goes on to note, "Once they get the government to stop predator control, the sheep industry will be in real trouble."[19]

Weed, however, doesn't see the battle in such stark terms. "I think we need to find a way to co-exist with the native species on the land," says Weed. "We consider ourselves to be both ranchers and environmentalists. I don't see an inherent conflict there."[20] She sees Predator Friendly as the way to do that. According to Weed, "The idea of marketing Predator Friendly Wool is a positive alternative to the wars being waged between the environmental and agricultural communities."[21]

Interest in the predator friendly concept is growing, too. Visitors from Tibet and Croatia have stopped by Weed's operation to explore the possibility of a predator friendly approach in their countries. Numerous letters and calls have come in from other interested parties around the world. "We have corresponded with a woman in Namibia who is training cattle ranchers there to use guard dogs to help protect the cheetah. A writer in Norway recently completed an article about the predator friendly idea in hopes of changing hostility to wolves in Scandinavia. A museum curator in Spain has asked to include predator friendly information in a textile exhibit in Barcelona," writes Weed.[22]

But perhaps more impressive is the interest in Weed's efforts closer to home. She explains that the hostility of some sheep ranchers to the idea is not universal. Weed remarks, "Many growers have called and written to me saying that do not feel represented by the stockgrower associations and they are intrigued with alternative approaches."[23]

Green Markets and Branding

Predator Friendly Wool's success relies on brand identification and labeling to capture extra revenue from consumers with a green demand. Building a brand name with a reputation for ecological friendliness creates a cachet that allows the producer to sell more product or to sell it for a higher price. Predator Friendly Wool is doing the same thing Fortune 500 companies such as Nike, Sony, or Pepsi-Cola have done. Consumers associate Nike with quality shoes, Sony with quality electronics, and Pepsi-Cola with a certain taste and quality in soda. Labeling provides independent certification that the product is, in fact, what it says it is: predator friendly. Becky Weed and her cooperative are working to get Predator Friendly associated with wool grown in an environment safe for predators.

Consumers who purchase Predator Friendly do so because they trust their purchase has reduced killing of wild species. An outside certifier verifies that claim for consumers, but the cooperative has a strong incentive not to cheat even without certification. If Predator Friendly growers did employ lethal methods on predators and the controversy was uncovered, one article in a newspaper or a report on a radio station could eliminate all of the benefits that the growers receive from consumers who pay their premium based on the wool's brand reputation. In the same way, Pepsi-Cola's brand would suffer if the company failed to exercise quality control.

Not too far from Becky Weed's operations, cattle producers in Montana's Madison River Valley are working to build a brand name of their own with Conservation Beef. The ranchmen's logo of an inverted C with a B hanging down from the C's back hails to the roots of the term branding. As far back as the medieval ages, hot irons were used for marking prisoners and slaves in case they escaped and for placing a mark of shame upon members of society who had committed acts deemed unsuitable. The end of chattel slavery and the advancement of civilization ended the barbaric days of marking humans with hot irons.

On the open range of the western United States, however, the use of hot irons for marking livestock as property kept branding alive. Cattle ranchers on the open range of the western United States used hot irons to sear combinations of diamonds, circles, letters, or other symbols onto the hides of their livestock. These symbols, the rancher's brand, denoted to whom the livestock belonged. It proved useful when running cattle with other ranchers or when identifying cattle stolen by cattle rustlers. As a side, it also proved useful in identifying successful ranchers who raised quality beef. A brand name could be associated with the quality of a product. Conservation Beef brings the past and the present together with its brand name.

Brian Kahn, a former leader of the Nature Conservancy of Montana, came up with the idea for Conservation Beef in 1992. As Kahn says, "I kept thinking we needed a way to get more resources to agricultural producers who were good stewards and who were interested in staying in agriculture."[24] He succeeded in forming Conservation Beef as a limited liability company in 1999 with the Nature Conservancy as a partner. Conservation Beef works with the Nature Conservancy, but its product is not sold with the Nature Conservancy name and logo. Therefore the Nature Conservancy's brand name is not on the line, nor do the producers for Conservation Beef receive the benefits of the Nature Conservancy name. Thus, the group is working to build its own brand recognition among consumers.

Ranchers who join Conservation Beef must develop an environmental management plan specific for their ranch. Plans include making enhancements for soil and water quality, wildlife habitat, and open space. For instance, the Sun Ranch works to protect riparian areas by drilling stock wells. This keeps cattle away from stream banks as they get their daily drink from the well instead of traipsing down through the erosion-prone soil near the property's many freshwater streams. Sun Ranch also lets down its fencing in the fall months so that elk migrating through the area can pass through without difficulty.[25]

Rangeland scientists, a board of ranchers, and a separate board approved by the Nature Conservancy review each ranch plan. Ranchers approved for the program are then eligible to sell their cattle to Conservation Beef, which in turn markets the beef to restaurants as well as to consumers through the Internet and direct mailings. To sell its product direct, Conservation Beef buys yearlings. It doesn't purchase calves in the fall with the responsibility of feeding those calves through the winter until they are large enough for slaughter. To deal with the cash flow problems that arise for ranchers who must wait until spring to sell, Conservation Beef gives ranchers a $400 down payment for each animal in the fall.

When the yearlings are finally sold, the rancher then receives 15¢ per pound above the current market price from Conservation Beef minus the down payment.[26] The organization makes up the 15¢ difference with the premium above market price that it receives from restaurants and consumers who are willing to pay more to ensure their beef is raised with the environment in mind.

How significant is the 15 cents above market price for ranchers? Assume a reasonable market price of 70¢ per pound on a 1000-pound yearling would bring $700 for a producer. Yet, someone enrolled in Conservation Beef could get paid 85¢ per pound with the brand name's 15-cent premium. This would bring the producer $850, a 21 percent increase in the overall revenue. In years where the market prices were depressed below 70¢, the percentage gain from the Conservation Beef premium would be even higher. Still, not all of that is strict profit. It must be remembered that the enhancements made to protect the environment cost the ranchers who must balance that cost versus the possible gains of selling under the brand.

One focus of the Conservation Beef program has been protecting open space. It accomplishes that goal in two ways. It succeeds first by virtue of its existence. Helping ranchers to obtain higher returns on their beef increases their likelihood of remaining in business, which means they won't sell the ranch into a subdivision or convert the property into a bunch of little ranchettes. But Conservation Beef goes a step further. Ranchers who want to continue with Conservation Beef must agree to a Full Stewardship Plan in their second year with specific commitments to open space. These commitments can be met by selling a permanent conservation easement, selling a term easement where they agree not to subdivide portions of the ranch for at least seven years or, if they do subdivide, they can give a portion of their income to a local land trust.[27] The land trust will then use the money to buy development rights in the area and retire them with a conservation easement.

In 2000, Conservation Beef purchased 149 calves, but Kahn has his sights set on much higher goals. He hopes within five years the operation will be annually purchasing between 10,000 and 30,000 head of cattle.[28] While Kahn's goals seem lofty, the operation seems to have the necessary backing to grow substantially.

Creating a brand from scratch can be a difficult process. It requires savvy marketing and an investment to get consumers to recognize the brand name—and be willing to pay more for it. Recognition of names such as Microsoft, Hormel, and Orvis is not built over night. For many farmers and ranchers, this route may seem a long row to hoe.

An alternative to building one's own brand is to go into business with an already established brand. In New Mexico and Arizona, several ranchers have struck a deal with Defenders of Wildlife where they directly employ Defenders' brand name and logo as part of the "Wolf Country Beef" seal of approval. In return for the Wolf Country seal, ranchers sign an agreement with Defenders of Wildlife that stipulates a number of provisions. The ranchers agree to donate one percent of their total proceeds to the Defenders' Wolf Compensation Trust, which pays ranchers for cattle and sheep losses caused by wolves. They agree to use their "best efforts to refrain from using non-selective lethal predator control measures, including but not limited to use of M44s, leg-hold traps, snares, and aerial gunning."[29] Lethal means of predator control are only allowed as a last resort. The ranchers also commit to help with reintroduction and monitoring of wolves on their property. Finally, enrollees agree to certain stipulations about the use of the U.S. Department of Agriculture's Animal Damage Control Services.[30]

In addition to letting the rancher use the Defenders' name and its logo, the program agrees to assist the ranchers in marketing their product and pays for Wolf Country seals and press releases. Still, it is the Defenders' name that has the potential to make consumers choose one brand of beef over another.

Twenty-seven year old Martin Heinrich purchases Wolf Country beef from his local co-op. "I really like the taste of red meat, but I spend a lot of time on public lands, and I'm not happy with the practices of ranchers on public lands in the Southwest." He thinks Wolf Country Beef is "a good way for people to express their support for wolf re-introduction."[31]

The Wolf Country beef ranchers benefit from the fact that Defenders of Wildlife bears a name many environmental consumers already recognize. Founded in 1947, Defenders has created a name and logo with customer recognition, and therefore great value in certain consumer circles because of its reputation for protecting wild species. While the Defenders' brand name gives the project a boost in its chances of success, it also puts the organization at greater risk for harm. If a rancher displaying the seal were caught killing wolves indiscriminately, that could hurt Defenders' reputation.

Arizona ranchers Will and Jan Holder enrolled in the Defenders' program to add a second brand name to the one they have already developed. Along with thirty other ranchers, the Holders have worked to establish Ervin's Natural Beef as the beef for green-conscious consumers of the southwestern U.S. The Holders run their operation as a "benevolent dictatorship" to ensure a reliable supply of quality beef for their customers.[32] In return, their fellow ranchers return profits guaranteed to clear $100 on every animal sold with the Ervin's name.

Producers of Ervin's Natural Beef are not allowed use of pesticides, insecticides, or artificial growth hormones. They also must obey Defenders' rules when it comes to predator control in order for Ervin's to use the Wolf Country beef seal. "Through non-lethal, pro-active management practices such as day herding, vigilance, and using guard animals, we can minimize loss to predators and make the few animals we do lose easier to accept economically," believes Holder.[33] The extra income from consumers who like Ervin's methods help make the few animals lost easier to accept as well. In addition, some growers are out for more than a profit. They consider themselves environmentalists and enjoy seeing predators around as part of the ecosystem.

Still, many consumers will not trust businesses, agricultural or otherwise, to patrol themselves simply to protect a brand name. For those consumers, producers can take another route to ensure customer trust and the increased revenue of green markets. This route entails product labeling. Branding products and labeling products differs because in the latter, standards for compliance are set for producers and enforced by an independent third party. To qualify for the label, the producers must meet the standards. Deborah Kane works with a labeling organization in the Pacific Northwest called The Food Alliance. For Kane, "The only drawback of a brand is that it's a self-endorsement."[34]

Whereas the reputation received through branding is self-enforced by the producer, the reputation from labeling is enforced by the independent third party: the labeling organization.

Green Markets and Labeling

The Salmon-Safe operation mentioned at the beginning of this chapter and the Predator Friendly cooperative that Becky Weeds joined are labeling organizations. Farmers or ranchers who hope to use the Salmon-Safe label must meet strict criteria, which are monitored and enforced by a third party certifier. An independent agency hired by Salmon-Safe inspects an operation using the criteria set by Salmon-Safe's guidelines.[35] In addition to traditional agriculture, Salmon-Safe certifies winegrowers in the region. After spending two years developing its standards, Salmon-Safe was launched in 1995 at the Sokol Blosser Winery.[36] By 1999, products labeled Salmon-Safe were being promoted in over 200 natural food stores and supermarkets, including a large chain of Fred Meyer stores.[37]

While Salmon-Safe is concerned with criteria that affect the health of fish in the Pacific Northwest, a program in the Midwest run by the World Wildlife Fund (WWF) focuses its efforts on reducing pesticides. Starting November 1 of 2001, 15 Wisconsin potato farmers and their 4,000 acres became certified by an independent auditor as eligible to employ the well-known panda bear logo of the WWF for selling their potatoes.[38] The farmers do not have to pay any licensing fee, they only have to meet pesticide reduction standards set by the WWF and verified by an Oakland, California auditor.[39]

With the help of the University of Wisconsin at Madison, farmers interested in using the WWF logo adopt practices including "basing their pesticide spraying schedules on observations of pests in the field rather than on the calendar; adopting spot-spraying techniques; and using beneficial insects to eat harmful bugs."[40] The World Wildlife Fund and the Wisconsin Potato and Vegetable Growers Association (WPVGA) began working together in 1996. The WPVGA represents over 250 potato farmers who raise 80,000 acres of potatoes in the nation's third largest potato-growing state. The WPVGA reduced 11 targeted pesticides by 500,000 pounds in three years, but it took nearly five years to get the environmentalists and the potato farmers to agree on a standard for shared use of the WWF logo.[41]

Jeff Wyman is a potato entomologist with the University of Wisconsin. He is excited about the new opportunity that the WWF logo offers for potato growers. According to Wyman, the potato growers "have worked closely with UW research and extension specialists since the 1980s to build the foundation for this unique program and they lead the nation in adoption of innovative strategies to reduce pesticide reliance. Economic recognition in the marketplace is long over due."[42] The WWF and the potato growers are working on a modest presumption of an extra 50¢ per hundredweight for the labeled potatoes.

The program focuses on fresh market potatoes rather than processing potatoes. To that end, 17 growers producing 9,000 acres of Wisconsin potatoes initially vied for the label.[43] But only 15 of the farmers and half the acreage qualified under the program's rigid standards. Jeff Dlott, a board member for one of the program's partners, thinks that's a good thing. "It shows that the standards weren't too easy to meet."[44]

Still, the program hopes those who failed to qualify and others who did not apply will one day meet the standards. Deana Sexson, a coordinator for the project, says, "Our goal is around 7,000 acres out of 25,000 fresh market potato acres in Wisconsin."[45] Once that goal is reached, the organization can move on from there. It is looking at adding vegetable growers in Florida and grape growers in California among others.[46]

While the program sounds like a branding operation, as it is using the WWF brand name, it more closely resembles labeling because of the independent, third party certification. The third-party certification is key for the WWF, which does not wish to risk its brand name, but still feels a need to provide an incentive for farmers to do the right thing.[47] In effect, the WWF is brand sharing. The organization shares its brand name, logo, and environmental reputation with the Wisconsin potato growers.

The WWF does bear a risk with the program beyond making sure that potato growers are reducing pesticides. If, for instance, a customer buys a bad bag of potatoes that does not taste good when cooked or has some other bad experience with the potatoes, it could reflect badly on the WWF's reputation with that consumer, even if the WWF has done nothing wrong.

Still, the program has good reason to believe it will make more customers happy than it upsets. Unlike many of the labeling and branding operations in this chapter, the WWF program has a more direct link to the consumer. It anticipates customers will buy the WWF-branded potatoes not just for the general well being of the planet, but for personal well being from lowered pesticide consumption. Sarah Lynch of the WWF points out, "As we looked into the eco labeling issue—we realized that industry-wide changes in pesticide toxicity were not what the consumer was really concerned about. They were worried about individual potatoes in a bag that they were putting in their shopping cart. They wanted to know about the potatoes in that bag."[48] The underlying assumption is that consumers will pay more as they see themselves directly impacted by the potatoes in the bags they buy.

In December of 1998, a new milk hit the shelves in the Fresh Fields grocery stores of Pennsylvania, Maryland, northern Virginia, and Washington, DC. Part of a one-year pilot program run by the Environmental Quality Initiative (EQI), the milk sent one nickel of every purchase to a stewardship fund. The fund paid a premium to dairy farmers who kept cows from riparian areas, managed their barnyard runoff from manure, protected well water, and were careful in storage and handling of pesticides.[49] Farmers were required to meet predetermined standards to qualify for part of the premium. The program was initially offered to 24 dairy farms in Pennsylvania, the nation's fourth largest milk producing state. It employed the University of Pennsylvania to evaluate the farms.

Lori Sandman works as executive director for the EQI. According to her, the project came about when, "We got a group of farmers together and asked them what we could do to encourage environmental practices. One answer was money to get the job done and the other was technical information."[50]

Unlike many labeling and branding programs, the product being sold didn't necessarily come from the farmers who were receiving the premium on the price. Rather, the five-cent premium charged on each half gallon of milk went into a fund. The dairy farmers who then qualified under the certification program were supposed to be paid 50¢ for every 100 pounds of milk they shipped. The test program, however, was not as successful as hoped for a couple reasons.

First, retailers were not convinced Chesapeake Milk would sell at a premium. Second, consumer confusion about how the milk related to popular organic milk hurt consumer demand. Finally, the project could not control shelf price, and some stores were selling the product at 60¢ over the price of a traditional half-gallon. The EQI felt this turned off consumers who knew that only a nickel extra was going to the farmers.[51]

Due to lower than expected sales, the program managed to pay farmers for participating, but it was not able to initiate its cost-sharing program for borderline dairy farms. With the cost-share program, the group hoped to help farmers who weren't quite meeting the standards to afford the necessary improvements for enrollment. For instance, money from the program could help a farmer build stream-bank fencing or a waterway crossing. Still, the program succeeded in showing that there was a market for a labeled product. Learning from the mistakes of the past, EQI is now working to launch a similar project in New York State for 2002.

While the project had mixed results on the marketing end, environmental improvement by the dairy farmers was all positive. "Even farmers who didn't qualify for the program ended up making changes," noted one of the on-site evaluators. "Farmers are concerned about the environment and realize these practices can make business sense as well."[52]

Perhaps one of the most successful labels on both the market and environment side is that of Portland, Oregon's The Food Alliance. Began in 1994 with a grant from the W.K. Kellogg Foundation, The Food Alliance had grown to include 60 farmers and ranchers among its enrollees by early 2001.[53] In March of 2001, Environmental Defense developed a strategic partnership with The Food Alliance to help promote the label. The partnership hopes to move beyond stores in Oregon and Washington into the national arena. According to Zach Willey, a senior economist with Environmental Defense:

> Giving consumers the ability to choose food according to production methods creates powerful new incentives to protect the environment. Farmers and ranchers face a recurring squeeze between low market prices and rising costs of production. Food Alliance labeling gives them a way to maintain and even expand market share, in some cases at better market prices. At the same time, consumers can't vote with their pocketbooks unless they know what they're buying. Food Alliance labels will provide that information.[54]

A farmer or rancher who wishes to qualify for The Food Alliance certification fills out an application and sends in a $300 application fee. Within four to six weeks, the property will be inspected and assessed on a number of criteria including pesticide use, irrigation practices, protection of riparian areas, treatment of workers, and numerous other crop and non-crop specific attributes.[55] If the site scores 70 percent or better on all of the different criteria, it is eligible for the label.[56] If the applicant is approved, they must then pay an annual fee based on their gross sales ranging from $250 for those taking in less than $25,000 annually to over $5000 for those with $1 million plus in sales. The fee is paid at the end of the year. In addition, the agrarian must submit an annual farm plan detailing changes in operations and improvements. The Food Alliance reserves the right to make two random and unannounced site visits each year.[57] If an applicant is rejected, they are given an analysis of what needs to be done to meet the standards. Once these changes have been made, the farmer or rancher is free to apply again.

The Food Alliance works to achieve a number of goals with its labeling organization. These include protecting and enhancing water and soil resources, reducing pesticides, conserving wildlife habitat, and providing safe working conditions for farm laborers.[58] In return for helping the organization with its goals, farmers and ranchers benefit from networking, increased market access and market share, on-farm research, sales support and advertising, marketplace education, public recognition and goodwill, credibility through third-party recognition, and customer loyalty.[59] All of these items help the farmers and ranchers to increase their bottom line. As Larry Thompson of Boring, Oregon, lauded, "I noticed increased movement two weeks after I won The Food Alliance's approval and began using their logo."[60]

The key distinction between labeling and branding is that by bringing in a third-party, independent auditor to certify whether standards are met or not, the labeling organization adds additional consumer confidence that operations are legitimate. Safe in the knowledge that the third-party verifier has no incentive to cheat in its auditing process for extra profits can help ease a consumer's mind that the label they are paying more to buy is indeed doing what it says it does. While there are many reasons for a brand name not to cheat or deceive its patrons, some consumers simply don't trust a seller when cheating could make a quick buck. Still, in the long run, cheaters in the marketplace never prosper; their reputation is for naught.

It is interesting to note that while the WWF potato program uses its brand name, it is in fact instituting a labeling program, and Defenders' of Wildlife Wolf Country beef purports to be a label but is in fact a brand name. An organization can, of course, do both. Becky Weed is building a brand name for her line of predator friendly wool products, while at the same time being certified by a third party to verify her predator friendly status.

Benefits of Green Markets

With the evolution of agriculture and the demands of an increasingly environment-conscious public, the marketplace is rewarding those who combine quality food with production methods protective of the environment. Higher prices, bigger market share, and good public relations are the major benefits on the agricultural business side. In addition, the ecological agrarians enjoy the added benefits of a cleaner, healthier environment and a feeling of pride when they kick back in their recliner at the end of the day. Consumers benefit from the cleaner environment and a feeling of good conscience in their purchased product. Finally, labeling organizations and environmental groups benefit in the same way as the eco-entrepreneur with financial gains and, in the case of environmental non-profits, from forwarding their conservation goals. As this book is about the rise of the ecological agrarian, we will focus on the benefits to this group before examining the benefits to consumers and labeling organizations.

Bonnie Chasteen works as the program director with the Corporation for the Northern Rockies in Livingston, Montana. This small nonprofit works to help agriculture find niche markets like the green markets discussed above. It helped establish the Predator Friendly Wool cooperative that includes Becky Weed as a member. The Corporation has been working to establish agrarians in green markets for over half a dozen years now, giving it valuable knowledge of what works and what does not. Chasteen provides a useful breakdown of the benefits that green markets provide agrarians. The first two benefits pave the way for the latter two.

1. They help restore producers' blackened image from land ravager to land steward.
2. They give producers opportunities for positive press and media.
3. They help producers gain access to new markets and increase market share.
4. They help producers get higher prices for their products.[61]

They help restore producers' blackened image from land ravager to land steward. One effect of the growing demand for environmental amenities is that environmental groups can rally citizens and politicians against agriculture for the harms of the past as well as those that still exist today. Through the political process, they can make business more difficult for agrarians by pushing for regulatory measures to deal with perceived environmental harms.[62] Agrarians can avoid these fights by improving their perceived reputation in the general public. Engaging in production that enhances the environment lets them take the high ground. This lowers the likelihood of conflict with environmental groups and, if there is conflict, it improves the agrarians' ability to defend themselves.

New Mexico rancher Jim Winder is enrolled in Defenders of Wildlife's "Wolf Country beef" program and grazes cattle on Bureau of Land Management (BLM) lands. He sees green markets as a wonderful opportunity to reestablish an agrarian reputation of land stewardship. "I want a family to drive by my BLM allotment and tell their kids—this is your public land and that guy over there in that house takes care of it. That's the way it ought to be—then we wouldn't have lawsuits and conflicts."[63]

They give producers opportunities for positive press and media. Becky Weed's operation is the kingpin of acquiring free press. She has been featured in publications like the *Christian Science Monitor, New York Times, Time* magazine, and *High Country News,* which reach millions of readers and potential customers. If only a fraction of those readers find inspiration to seek out her website at www.lambandwool.com, Weed can get a significant hike in her sales. As her business grows, word of mouth will help the product grow too, but those free media hits are a nice bonus.

Journalists enjoy telling feel good stories about businesses doing good for the planet and for themselves. Agrarians who enter green markets are protagonists for coming issues of the *Wall Street Journal, Newsweek, Reader's Digest,* and *USA Today.* Some labeling organizations like The Food Alliance actively work to get spots in local television, radio, and print media. They know that many of their strongest customers will be local buyers. Defenders of Wildlife, in fact, launched Wolf Country beef with the purpose of creating a marketing and media blitz to extend their brand and forward their environmental goals.

Still, media exposure isn't everything according to Weed. "The media coverage has certainly helped us get the word out, but most sales come more from direct relationships than AP wires."[64]

They help producers gain access to new markets and increase market share. Many agrarians do not need higher prices as much as they just need people to buy their product. It doesn't matter what a pound of apples sells for in the market; an orchard and its owners do not benefit unless people are buying their apples. Labels and brand names can be the difference between a customer choosing one product over another. As seen in the Michigan State survey, 72 percent of the apple buyers claimed they would choose eco-labeled apples over regular apples if both were the same price. The eco-labeled apple producer in that case is not receiving a premium on each bag sold in comparison to her competitors, but she might end up selling a lot more bags than they do. More bags sold at a profit equals more money made.

Lowell Catlett works as an agricultural economics professor at New Mexico State University. In September of 2000, he spoke to a number of farmers and ranchers on the new markets arriving in a society blessed with disposable income. "You need to learn to differentiate your products when selling to different generations of clients," Catlett advised. "Each generation has different needs."[65]

Sometimes a labeling organization can open up a market that didn't even exist before. Gary Wells grows apples in Hood River, Oregon. He regularly sells them to Asia, but when the Asian market collapsed at the end of the 1990s, he found himself stuck with 200,000 boxes of apples. The Food Alliance stepped in to broker a deal for Wells with one of the largest chain stores in the area, despite the fact that the market for apples was glutted.[66] As a member of their family, The Food Alliance wanted to do everything it could to keep Wells in business.

They help producers get higher prices for their products. Labeling and branding connect consumers who are willing to pay more for environmentally friendly goods with agriculturalists who are happy to earn the extra revenue. Becky Weed is getting two times market price for wool. Conservation Beef producers make 15¢ per pound above market. One producer of frozen fruits and vegetables garnered 15 percent more than competitors thanks to labeling by The Food Alliance.[67] Consumers paid a nickel more per half-gallon in order to buy Chesapeake Milk.

🌾 🌾 🌾

We've discussed the major benefits for the agrarian producers, but as mentioned, there are benefits for the consumers and the labeling organization as well. Consumers choose to pay the higher prices because they get a good feeling from helping out the environment. In addition, they benefit from the better environment that they are paying to support. The labeling groups stand to benefit from the business opportunity of charging fees to members to support labeling activities. Those who share their brand name like Defenders of Wildlife can benefit from tying their brand name to a product that increases the brand's reputation. Finally, most labeling groups are succeeding by forwarding the environmental goals of their organization.

Another important component of green markets from the standpoint of environmental organizations and labelers is keeping farmers and ranchers that share similar values in business. By adding to the financial opportu-

nities for agrarians interested in improving the environmental management of their operations, environmental groups give these agrarians a
hand up in the marketplace. Offering eco-friendly farmers and ranchers a
better chance of survival increases the chance that land will be managed
with green goals in mind.

It is clear that benefits can accrue for producing a product with a reputation of good environmental practices. One must be careful, however,
to realize that there are costs and dangers to entering the world of green
marketing. Assuming that being a friend to the environment is enough to
sell a product can prove fatal for any business. David Riggle edits *In
Business*, a magazine for eco-entrepreneurs. He finds that too often eco-
entrepreneurs make this mistake. "If it's food, it has to taste good. If it's
an energy saver, it has to work. It has to be a good product," reminds
Riggle.[68]

Agriculture's current evolution does not mean that producers can
abandon the achieved goal of feeding people a quality meal for the new
goal of environment friendly production. Being green is not enough to sell
a product. The product has to be useful and high quality from the start. It
is the product that keeps people coming back for more. The product provides the cake. The production process is only the icing.

"Green marketing is like any other business," cautions Becky Weed,
"you must have a product that people really like and think is high quality.
It just has the added challenge of seeking a premium price."[69]

In taking the next step up the evolutionary ladder of selling food,
farmers and ranchers must not abandon the preceding steps. People buy
food for sustenance first. Then they buy it because it tastes good. They
will not continue to purchase beef, bananas, or even Becky Weed blankets
solely because the products are made with environmentally-friendly
means of production. The utility and quality criteria must be met and
continue to be met under the new means of production. As Deborah Kane
of The Food Alliance cautions, "The label can open up markets, but you
really have to start with an outstanding product."[70]

Brian Kahn, who provides the entrepreneurial spirit behind Conservation Beef, understands this concept as well. "People won't buy this beef
twice because it's Conservation Beef. They will if it's a superior product,"
comments Kahn. Remembering that his endeavor is first and foremost a
business provides Kahn with the grounding that gives his product a real
chance of succeeding. Making the mistaken assumption that a green label

or brand name will sell a product regardless of quality is only one of many challenges and dangers faced when an entrepreneur reaches for the benefits of green markets.

Challenges and Dangers of Green Markets

Green marketing isn't as easy as one, two, three. There are many challenges and costs for the farmer or rancher who enters a green market, the labeling and branding organizations that certify them, and the customers who must pay the higher price to make it worthwhile for agriculture to go green. These challenges are not insurmountable, but there may be cases where it is not worth the effort to obtain a green label.

Challenges for Agrarians

Farmers and ranchers are cautious by nature. Living from one year to the next tends to make them risk averse. One bad year could force the agrarian to sell off land to pay their bills or get a job in town. In a worst case scenario, they may end up having to sell the whole property. This is the ultimate fear for those who have had a spread of property in their family for generations. Like Scarlett O'Hara in *Gone with the Wind*, the land is their final identity, something they can always fall back on when other things go awry. Change entails risk, and many farmers are not willing to incur risk unless they have reached the end of the line with nowhere else to turn.[71]

Green labels and branding entail a number of risks for the farmer or rancher. One major risk is getting involved with a labeling organization or borrowing a brand from a group that does not share a comparable philosophy or land ethic. This can be embarrassing for the landowner and the landowner's product. It can prove dangerous to the long-term viability of their business. For instance, the American Humane Association created a program to certify cattle, pork, and poultry grown in a humane manner. Those who qualify for the program are free to label their products with the "Free Farmed" label.[72] The program has laudable goals and does not appear to be a significant danger to ranchers. But if the program were run by The Humane Society of the United States (HSUS), ranchers would be wise to be wary of entry. With its three R's program for choosing a humane diet, the HSUS advocates that consumers "Replace animal-based foods with foods that don't come from animals."[73] By supporting such a program and group, ranchers would be endangering their own lifestyle. They would be supporting a group that opposed them. To avoid this risk, farmers and ranchers need to ask what the labeling organization or owner of the brand that they are sharing has as final goals. What are the organization's values? Is it trying to end the agrarian lifestyle or merely change aspects of it?

Related to the risk of establishing a comparable philosophy, it would be helpful for an agrarian to know who funds the labeling organization and what their philosophy is for the same reasons as those mentioned above. But this isn't the only reason to know who is funding the labeling organization. If an agrarian is going to make the effort to invest in changes in the way they do business, they should be sure that the label or brand name will be around long enough for them to benefit from their investments. The Wolf Country Beef program run by Defenders of Wildlife was never intended to be a long-term program. Rather, Defenders used the program to show green marketing could be done and to get publicity for the issue of wolf reintroduction.[74] If a labeling organization has unstable funding or lacks a long-term plan, the agrarian might be better off avoiding the investment in new operations. In the case of Wolf Country Beef, it is clear that Defenders will be around for some time. The question is how long they will support the Wolf Country Beef program. Farmers and ranchers need to ask what the labeling or branding organization's commitment is and what is the commitment of their funding? Agrarians must decide if the investment made to enroll in the program can be recouped.

Another question that must be considered when trying to cover investments is whether the standards for keeping the label will remain the same or whether they are subject to change. If the standards can easily be changed, agrarians risk investing in changes to their operations and then finding the bar raised, making them ineligible for the label without further costly changes. Making sure "the group's standards are as clear and measurable as possible and that it makes communicating these standards (your standards) a priority," comments Chasteen can reduce part of this risk.[75] In addition, the agrarian should try for a contract guaranteeing their use of the label at least long enough to cover their investment costs or a contract that clearly lays out the conditions under which standards can be changed.

Another risk for agrarians getting involved in labeling lies on a larger scale than their individual involvement in labeling. By promoting labeling programs, there is a risk that the government will step in and use the labels as an excuse for mandatory labeling. Mandatory labels take away the advantage that farmers and ranchers achieved by attaining the voluntary label. The agrarian is left with the costs of changing their operations, but no benefits. Mandatory labels can eliminate the niche that the agrarian was marketing in because everyone is now forced to meet the standards of that niche. If everyone is raising Wolf Country Beef, there is no advantage for the Wolf Country Beef rancher over his competitors. The consumer does not need to go out of their way to buy from the branded or labeled agrarian. Government labeling also tends to stop advances by freezing into law the politically acceptable solution of today. While herbicide use may be the important labeling item this year, soil erosion could be important next year. Government labels are slow to adapt to such changes.[76] In addition, when labeling is made mandatory, those selling labeled products do not have the option of leaving the program if its standards become too costly. Their only option is to get out of business.

Due to the number of risks involved and the risk-averse nature of traditional agrarians, it is not surprising that many of the early adopters of ecological branding and labeling are farmers and ranchers who can afford losses or agrarians with less of a tie to the land such as those using farming or ranching as a hobby or retirement career. For instance, the owner of Willamette Farms is a successful lawyer. He can afford the possibility that his agrarian endeavors will fail. Becky Weed and her husband were both professionals (a geologist and an engineer) who decided to get into ranching. They were environmentalists first and then agrarians. Those whose whole lives are tied up in the land and who come from several generations of farmers and ranchers may have a harder time betting the farm on a new venture than so-called Johnny-come-latelies.

These early adopters are necessary to get ideas off the ground, but if green markets are to grow, traditional agrarians will have to get involved. The trick for those who cannot afford to lose is balancing between entering too early and getting left out in the cold by entering too late. The agrarian must weigh whether the benefits outweigh the costs and whether green markets will lead to a higher return than continuing with the status quo. They must decide when the risks are reasonable to take for an increased return.

There are risks to staying on the sidelines as well. There is only room for so many brands and labels in the consumer's shopping cart. Those who begin branding early will have an advantage as their brands achieve recognition with customers. Once customers associate with a certain brand or label, it can be quite difficult for competing brands and labels to gain their attention.

In addition to all of these risks are the direct costs and challenges of meeting the standards of labelers or building a brand name. For instance, it costs money to have the assessment done to qualify for a program. As noted above, acquiring the Salmon-Safe label entails an assessment fee of between $400 and $500 on average. In addition, there are costs to making the improvements necessary for meeting label or brand standards. Some agrarians may already be engaged in the necessary practices, to lower pesticide use, or reduce the impact of grazing, but many will have to invest in costly procedures and capital improvements. If an agrarian or a group of agrarians is working to build a brand name like Conservation Beef, it can entail substantial advertising and marketing costs to build customer recognition. On top of these costs, agrarians may have to expend time and labor to meet the demands of green consumers.

The nature of today's agriculture can also make it difficult for certain commodities to seek a label or a brand. The wheat market is less conducive to direct marketing like beef or fruit where the producers can sell to the retailers direct. Instead, it is processed by large wholesalers where the link between the food produced and the food consumed is not clear. This makes it difficult to tie the production process and the consumer demand to one another. Challenges, however, are nothing more than opportunities for entrepreneurs as those who overcome the challenges are the ones who will reap the benefits. One member of the Salmon-Safe labeling group grows wheat and barley. He plans to sell his product direct to a brewery that wants to sell a specialty beer with the Salmon-Safe label. In addition, the growing market for organic goods has opened the food industry up through local food cooperatives.

Risks and costs do create barriers to entering the green market, but as Bonnie Chasteen writes, "If it was cheap and easy, everybody could play" and if everyone played, the extra profits would not be available.[77] The key for agrarians who want to succeed in green marketing is the same as in any other business; they must do their homework. Farmers and ranchers must consider the standards, the costs of meeting those standards, and the costs of assessing standard compliance. They must make sure standards are clear, measurable, and achievable so there is little room for fudging. They must ask whether the labeler's ideology is agreeable and whether the group works to market the label. If customers don't know about the label or the brand, what value is it?

Balancing the numerous costs with the potential benefits in the agrarians' decision-making process will distinguish between success and failure. If the higher prices received from a green product or label do not justify the expenses in time, money, and risks to acquire green status, agrarians should stay out of the program. But as the examples above indicate, there are ways for the agrarian to lower the risks of entering a green market and lucrative opportunities for those who succeed.

Challenges for Labelers and Branders

Challenges for labeling organizations parallel the challenges for agrarians. Organizational stability is an important consideration for the agrarian who opts for labeling, but it is important to the labeling organization as well. The goals they can achieve are limited by how long they are in business. Labeling organizations would be wise to have a business plan in place that is self-funding from fees within three to five years. Most labeling organizations seem to achieve charitable funding from foundations to run their operations for a few years, but after that, foundations expect labelers to make it on their own. If a labeling organization hasn't set up a solid business plan, it might not get funding in the first place. Even if it does, it may be out of business in a handful of years without a good, long-term strategy.

Another way that labelers can extend their efforts is by finding a group with similar goals to work with as a partner or as a mentor. The Food Alliance, for instance, is partnered with The Midwest Food Alliance in St. Paul, Minnesota.[78] This helps expand marketing opportunities for both organizations.

Just as the agrarians benefit from clear and measurable standards, the labeling organization benefits as well. Clear standards lower the opportunities for miscommunication that could lead to a bad reputation in the agricultural community or even legal action due to breach of contract. Specific standards reduce the difficulty of measuring success and ease the process of explaining to rejected applicants why they have failed to qualify.

Brand sharers like Defenders of Wildlife risk their name when they enter into agreements like Wolf Country Beef. If it is found that someone using the Wolf Country brand name is not, in fact, protecting wolves, Defenders of Wildlife could lose support from its membership and value for their brand reputation as an environmental defender. Thus, it is imperative that they be careful in choosing to whom they will lend their brand name. Labeling organizations bear the same risk. If someone using their label is not meeting their standards, they need to take immediate action to avoid bad press that could reduce the value of their label in the eyes of the consumer.

The branding or labeling organization also bears most of the cost of getting the brand name or label recognized. This can be the most expensive part of the process. While free press is available for unique, positive programs like labeling groups, selling the product to restaurateurs or supermarkets requires advertising costs, promotional events, and an able sales staff. Chefs and buying managers for stores often need to be sold before the customer can be sold. The Internet offers one medium to reach the customer directly, but the intermediary retailers are key to reaching a larger market.

Challenges for Consumers

Consumers are ultimately the ones who decide whether a green product succeeds or fails. They are the ones who select one product over another because of differing merits. They are the ones who pay higher prices for one brand name or label over another. In this way, they are the ones who pay the direct costs and provide the incentive for the agrarians and labeling organizations to do what they do.

Beyond digging into their pocketbook, consumers incur costs by going out of their way to find products or by requesting that a supermarket carry a certain line of items. In return for their efforts and their discretionary spending, the consumers get the benefit of knowing they are purchasing a product that makes them proud. This is where the challenges for consumers come into the picture.

Perhaps the biggest difficulty for consumers of environmentally-friendly products is knowing whether the product is indeed good for the environment. Decisions to raise organic food reduce pesticide use but at the same time, it may increase soil run-off and erosion while increasing the amount of land that must be cultivated for agriculture. For instance, DeGregori argues that "Modern no-tillage (or reduced or minimum tillage) agriculture using pesticides for weed and pest control conserves both soil and water better than its organic competitors."[79] To make decisions, consumers will have to do more research or trust third parties to provide the necessary information for them.

Consumers may have to choose between competing environmental goods too. Are they more worried about erosion and water quality or pesticide use? With the dynamics of our changing world and information increasing exponentially from new studies and scientific research, consumers' values won't remain constant. One day pollution from coal-fired energy plants could be the concern and the next it might be birds dying from the turbines that generate wind power. The ecological agrarian who hopes to stay in business must anticipate and react to the consumers changing goals for environmental production.

Often brand names won't mean anything, either. For instance, General Mills wrote a check for $115,000 to the Nature Conservancy in return for the right to use its oak-leaf logo on boxes of Nature Valley granola bars. The logo signifies no environmental benefits beyond those attained by the Nature Conservancy from the cash payment.[80] This may be enough for some consumers, but others will want a more tangible payoff. One place that consumers can go to check out what a specific eco-label means is www.eco-labels.org. This website catalogs different eco-labels, what they stand for, who certifies them, and so forth.

A final problem for consumers is too much information. In the organic business, Gardner notes that presently, "forty-four independent bodies— thirty-three private and eleven state—are responsible for setting organic food standards."[81] With a proliferation of labels, it becomes difficult to determine what to buy and what not to buy. This is often overcome by sticking with proven brands; hence, the benefits for agrarians of sharing a brand with an organization like the World Wildlife Fund or Defenders of Wildlife. Another option to help ease consumer worries is umbrella organizations that certify the labelers and sift through the information for consumers. Umbrella organizations work to ensure that labelers are meeting the claimed standards and reduce the number of labels about which a consumer really needs to be informed. The consumer can just look for the approval of the umbrella organization in the form of a label and then ignore the other labels. Supermarkets can work as filters for information, too. Stores like Fresh Fields in the Washington, D.C. area have grown by earning a reputation for buying goods that appeal to a certain consumer interested in food produced in a certain way.

The Key to Green Success

The challenges of entering into green market opportunities are many, but they are not insurmountable as this chapter's ecological agrarians demonstrate. For those who succeed in tackling the emerging agricultural markets of the future, the benefits include higher prices, more products sold, and continued pride in a job well done. The key to success is balancing the risks with the potential gains. A well-researched assessment of the pros and cons combined with a cautious weighing of potential risks is by no means an easy task, but staying in business never is—that's why the payoff is so rewarding.

Chasteen listened to a panel of ranchers discussing niche marketing at a December 1998 conference in Phoenix, Arizona. In hearing what they had to say, she found that

> Although these producers make environmental claims or carry an environmental seal of approval, they concede that they didn't brave a path into the relatively uncharted territory of niche marketing because of their ecological values. This conference has included a good deal of talk about learning the language of ecology to improve management and relationships with agencies and environmentalists, but the main thrust of the programs has been economics.[82]

As Anderson and Leal quoted one Montana rancher, "If it pays, it stays."[83] This is the ultimate challenge to green markets for agriculture. If there is no demand from consumers for environmental goods, they won't exist as a niche in the marketplace. If the product cannot maintain its quality when offering the green label, it will fail. In order for a business to be sustainable in the long run, it has to make a profit. It has to attain positive cash flow. Benefits must outweigh costs. The eco-entrepreneurs we met in this chapter are the pioneers who will make that happen.

3.

WASTE NOT,
WANT NOT

Waste not, want not. One person's trash is another person's treasure. These adages were in practice long before recycling became a catchphrase of the environmental movement. While environmentalism encourages recycling through regulatory edict, public education, and peer guilt, more effective recycling occurs because it is cost effective. Things like using the back of an envelope as scratch paper, passing hand-me-downs to a relative, or even saving leftovers for the next day's lunch are all forms of recycling that we practice every day. This recycling is done not to feel good about ourselves, but because it saves time, money, and effort.

Recycling occurs in agriculture for the same reasons. Farmers and ranchers have used manure from stock animals to fertilize crops for centuries, because it was in their best interest. As time marches on, technology, resource availability, and demands change. Changes have encouraged new forms of recycling while other forms have become less worthwhile. There is little point in recycling horse manure when commercial fertilizers are cheap and don't require collection and processing. This is no different from the handing down of clothes to relatives declining over time as clothing has become cheaper and incomes have risen.

In contrast to the decline of certain forms of recycling, the tales of beer, sewage, thirsty cows, and thirsty fields below focus on new forms of recycling that agriculture is engaging in as demands, resource availability, and technologies change. Reasonable recycling does not need government-mandated regulations or subsidies; it only needs entrepreneurs and ecological agrarians seeking to improve the bottom line.

Buying a Brew for Bessie

Beer is as American as baseball and apple pie. Each year, Americans pop the top on more than 5.6 billion gallons of beer.[1] Most of us can imagine where that 19.5 gallons per person ends up. Many of us do our fair share to help out the brewers every Fourth of July, New Year's Eve, and Friday Happy Hour, but few of us understand what it takes to make our favorite suds.

From large brewing companies to the craft brewer, the beer industry prides itself on its ingredients. Besides beautiful women, funny guys, and former athletes, the commercials of the beer industry emphasize the purity, flavor, and general superiority of their ingredients over their competitors'. Without the finest hops, barley, and yeast, beer is nothing more than bottled water.

The H_2O aside, grain is the main ingredient in beer. The industry uses more than 400 million tons of grains annually. Everything from rice to wheat is used to brew America's national drink, but barley is the grain of choice. Long before beer drinkers can imbibe, brewers must transform grains into delightful ales, lagers, and porters. The brewing process begins by grinding barley or other grains and immersing them in water. The mixture is boiled to extract the sugars and starches. The rich sugary brew is siphoned off to fermentation tanks. Left behind are heaps of wet, soggy grains.

Beer manufacturers are quite adept at disposing of the beer. They bottle, keg, or can it for sale to thirsty consumers. But getting rid of the mountains of wet grains, often called spent or brewers' grains, is not so easy. One option is to send the slop to landfills, but that is expensive. So the industry has applied the old adage and found that its trash is indeed the treasure of agrarians willing to pay good money for the unwanted piles of waste. From big to small, brewers partner with agrarians to reduce, reuse, and recycle spent grains.

Beer: A Booming Industry

Beer manufacturers recycle their leftovers into full-course meals for agriculture. They have created markets for the wet grain and related brewing by-products. By selling brewer's grain as livestock feed and brewer's yeast as a food supplement for dogs, the beer industry is saving millions of dollars in disposal fees. One of the factors driving these products has been the growth of the industry.

Beer manufacturers produced more than 4.5 million tons of spent grains in 2000 and that number is likely to increase in the coming years. The year 2000 also marked the U.S. brewing industry's fifth consecutive year of growth.[2] Beer shipments from brewers to wholesalers set a record, rising to 197.6 million barrels. With such continuous growth, firms in the competitive brewing industry need to become even more creative in disposing of grains and other brewing by-products.

Where some see the increased volume of spent grains as a problem, industry entrepreneurs such as George Wornson see opportunity. Wornson is the manager of Miller Brewing's by-products business. To him, "A brewery is really a beer factory and produces many by-products."[3] He believes that innovation requires looking at the income possibilities of these brewing leftovers.

Wornson helped Miller develop a corporate philosophy toward recycling and reducing waste. In 1990, Miller Brewing committed itself to reducing its use of landfill space by 25 percent per year. As the company realized the cost savings, the goal was ratcheted up to virtually eliminating use by 1995. To achieve this goal, the company had to find new uses for its grain waste. Wornson helped Miller implement a byproduct development program that reclaims material to be marketed to other outlets, thus generating income rather than paying to use landfills. This is where the agrarians came into play. The spent grains were marketed under the brand name *Barley's Best* and sold to farms and commercial bakeries as a fiber supplement.

Other brewers have also recognized the value remaining in their leftover mash. Anheuser-Busch keeps all of its spent grain out of landfills. Most ends up as food for milk cows. Cattle feedlots also buy the company's soggy grains, but the dairy industry continues to be the biggest customer. In 1999, Anheuser-Busch sold 1.76 million tons of spent grains to dairy farmers to feed more than 200,000 cows. By finding recycling markets for many byproducts, Anheuser-Busch breweries generate revenues and save nearly $1 million per year in landfill charges.[4]

"That's good for the animals that receive it because it's premium quality stuff; it's what we put into our beer, and at the same time we're able to make a profit selling the grains," commented Charles Poole, director of consumer affairs and environmental communication at Anheuser-Busch.[5]

One problem with the grains that breweries sell to farmers is that they are wet. Most farmers prefer dried grains because they are easier to store and handle. The fact that most grains arrive at the farm in a soggy form poses some limitations for the recycled grain market. Wet grains have a short shelf life and need to be consumed before they spoil. The grains can be stored for only two weeks before the smell could "gag a maggot," reports Mark Hissa, an Ohio dairy farmer who uses spent grains as feed.[6]

Wet grains are also expensive to transport and ship because of the added water weight. A study by Chandler and Associates of Dresden, Tennessee, found that the movement of spent grain in its wet form is limited to within 200 miles of the customer due to freight costs.[7] Consequently, the use of spent grains as feed tends to concentrate around areas served by major breweries.

During the early days of the wet grain industry, larger beer manufacturers installed drying facilities to make their grain more appealing to farmers. However, most major brewers are moving away from drying because it is expensive and energy intensive. At one time, Miller Brewing Company had drying facilities at nearly every brewery. According to Dan Dwyer, Miller's byproducts manager, only three of the company's breweries have dryers today. "Farms like dry grain, but there's an economic trade off because drying is expensive."[8]

The Nebraska Institute of Agriculture and Natural Resources (IANR) looked at the nutritional and economic benefits of feeding wet versus dry grain. The study found that feeding wet byproducts to livestock compared to dried grains yielded cumulative net economic benefits of $215 million in Nebraska from 1992 through 1999.[9] Annual net economic benefits of feeding wet byproducts grew from about $1 million in 1992 to an average of $43 million in recent years as new processing plants opened and more Nebraska feedlots fed wet byproducts.

The research was important in demonstrating the value of feeding wet grains instead of dry grains to animals. It showed wet grains were economical and performed nutritionally as well as, if not better than, than traditional corn rations. While most producers prefer dried grains, the IANR study also showed that drying actually reduces the nutritional value.[10]

Whether grain is wet or dry, major brewers have developed a variety of outlets for their spent grains. "We're always looking for new uses for our

spent grains," said Steve Rockhold of the Coors brewery.[11] Working to further expand its outlets, Coors started forming pellets with some of its
grain so that the product can be shipped internationally. The pellet-
shaped grain makes tasty bite-size morsels for livestock in foreign markets. The pellets are also easier to handle, ship, and store, allowing Coors
to move its spent grain beyond the traditional 200 mile market area.
Rockhold estimates that the largest growth in Coors spent grain sales has
come from overseas purchases.

A New Kind of Grain Company

While the spent grain market is now sophisticated, feeding brewers' waste
to livestock is as old as brewing itself. Up and down the East Coast of the
United States, hogs were frequently raised near liquor distilleries and
were fed on the mash.[12] During the nineteenth century, cows living on the
swill of local breweries produced most of New York City's milk.[13] According to Charlie Staff of the Distillers Grain Technology Council, Kentucky
moonshiners once fed corn mash waste to hogs as a way to hide incrimi-
nating evidence from revenuers.[14] Still, it's only been in the last 50 years
that spent grains developed into a viable commercial market. During that
time per capita beer consumption has risen from 12 to nearly 20 gallons
annually, with total production jumping by more than 20 percent.[15] More
beer production means more spent grain.

While the industry is delighted with the growth in beer sales, it has had
to grapple with grain disposal. Prior to the development of spent grain
markets, one common practice was to compost the grains. Composting is
a polite description. Most breweries piled the grains, leaving them to rot.
But the smell of decomposing grains was so intense that local residents
often complained.

Other beer manufacturers turned to agriculture for help. Some started
their own small feedlots, where the leftover grains were used to feed livestock. But most manufacturers called on local farmers to haul off the
mounds of grains. Yet, as the brewing industry grew, the volume of spent
grains was too large for local farmers to handle. The large brewers began
employing brokers or consolidators to market the spent grain. Companies
such as F. L. Emmert, a spent grain marketing firm, developed with the
brewing industry.

The roots of the F. L. Emmert Company date back to 1881, when
Frederick Emmert ran a saloon at the corner of Vine Street and Clifton

Avenue in Cincinnati. Several of Emmert's better customers were local German brewers. One of their constant complaints was that local farmers who were supposed to pick up the spent grain for their livestock were late or did not show up at all. Emmert, a born entrepreneur, offered to take all of the spent grain and resell it. The nineteenth-century grain recycling business of F. L. Emmert Company was born.

In the late 1920s and early 1930s, the company began to dry the spent grains. This drying process enabled the finished product to be shipped further distances and, more importantly, to be stored for long periods of time. At that time, F. L. Emmert Company was one of a handful of companies in the country drying spent brewer's grain. It began shipping dried brewer's grain to all parts of the country. At about the same time, prohibition was repealed and a number of breweries were either started or re-opened in Cincinnati. This gave Emmert a boost in supplies.

In the late 1930s, the company developed a proprietary process for adding wet cane molasses to the brewer's grain. Prior to this break-through, adding any significant amount of molasses to brewers grain resulted in a product that "set up," got hard, and was near impossible to handle. Molasso Malt, as the new product was named, allowed dairies to add highly nutritious brewers grain and the palatability of molasses to this ration in an easy-to-handle, economical way. It soon became an indispensable ingredient for dairy rations throughout the Midwest and was the company's best-selling product for the next twenty years.

The company expanded into other brewing byproducts over the years. Around 1960, it experimented with the spent yeast from the local breweries and after several years of development introduced a product known as BGY-35. Primarily sold to feed and pet food manufacturers in the eastern United States, the dried yeast opened up new markets for the company, boosting sales significantly. Using its knowledge of brewer's yeast, the company also developed nutritional supplement products for livestock, dogs, and cats.

Micro-brewing Industry Boom

The revival of local brews has renewed an old tradition between local farmers and breweries. Before companies like F. L. Emmert Co., farmers would show up each week and haul off as much spent grain as needed. They took small loads that would fit in their farm trucks, but the demand was enough to get rid of the spent grain. After World War II, however, consolidation occurred in the industry and small local breweries gave way to large commercial brewers like Anheuser-Busch and Miller. These companies produced an enormous amount of beer and more production meant more spent grain. They produced more spent grain than local farmers could handle.

While we have seen how the large brewers handled their waste, local farmers have again become vital for waste disposal for small microbreweries like Cleveland's Great Lakes Brewing Company. Great Lakes and six other breweries contract with Mark Hissa, who runs a dairy just outside of Cleveland. The dairy sends a dump truck into the city about five times a week, picking up mushy mixes of wheat, oats, and barley.

"Each cow gets a big shovelful in the morning and one at night," explains Hissa. The grain has lost much of its sugars, enzymes, and flavor in the brewing process, but according to Hissa, it contains enough protein to supplement a cow's regular rations of corn, dry grains, and hay. He notes, "The spent grain still has a lot of energy in it."[16]

The dairy makes its biggest haul from Great Lakes, Cleveland's first microbrewery. Great Lakes produces about 25 tons of spent grains per week, while each of Hissa's 130 dairy cows eats about 40 pounds a day. Hence, Hissa's hungry herd takes in all of Great Lakes' leftover mash and then some. Without the dairy, the brewery would have to pay a hauler to take the spent grains away for dumping in a landfill.

Cleveland breweries and dairy farms have a longstanding relationship. During the early part of the twentieth century, farmers all over the region drove horse and wagon to Cleveland to pick up spent grains from the city's breweries. In 1910, Cleveland had as many as 26 breweries. Sixty years later, only two were still in business and by the mid-1980s, there were none. With the big brewers dominating the industry, the traditional link between the farmers and the breweries died.

The Conway brothers opened Cleveland's first microbrewery in 1988. During the early days of operation, the brewery only utilized a half truckload of grain each week. That same year, dairy farms were struggling with a shortage of food supplements for their cattle due to an extended drought. Even a small amount of grain was helpful. As luck would have it, Mark Hissa's uncle stopped into the Conways' new brewpub for an evening pint and spotted barrels of spent grain stacked in the alley. He mentioned it to his nephew and shortly afterward a handshake agreement was struck between Hissa and Great Lakes. The brewery did not want money for the grain as long as the dairy agreed to pick up the grain on a regular basis.

The deal turned out to benefit both parties. The brewery's popularity grew and it started making more beer. "Pretty soon they were brewing enough for two pickup loads a week," recalls a joyful Hissa.[17] As other microbreweries opened, word of the dairy's free pickup service got around. With increased demand, Hissa had to buy a small dump truck to handle the volume.

The brewery's spent grains have provided the dairy with a reliable source of feed. "The cows gobble it up," said Jim Conway, co-owner of Great Lakes.[18] Other dairies have tried to get spent grains from Great Lakes, but a handshake and good service sealed the bond of an exclusive agreement between Hissa and the brewery. "We do a good service for them," says Hissa. "And they've been loyal to us."[19]

Beer-Drinking Cows

At first, it sort of sounds like the punch line to a bad joke—cows drinking beer. But it is an innovative solution to a real brewing problem—what do you do with beer that has outlived its shelf life? Until recently, the Canadian brewery Molson paid the city of Edmonton to dispose of stale outdated beer. Now that same beer is being served to cattle at a nearby ranch.

The idea of fattening bovines with beer isn't new. Kobe beef cattle in Japan are fed a special diet that includes a plenitude of pilsners. But a feed program designed specifically to save money is a new twist. Peter Rochefort, environmental specialist at Molson's Edmonton brewery, developed the idea to recycle the company's waste products. "The brewery used to dump all its stale and mislabeled beer into the sewer system," said Rochefort.[20] Up to five million bottles of beer were sent down the drain each year.

Not any more. Today, Molson's outdated brews are mixed with the regular cattle feed to create a kind of wet mash. Each cow gets a daily allotment of 10 pounds of beer (the equivalent of about 12 bottles) mixed with 40 pounds of feed. That might seem like a lot of beer to give a cow, but they don't get tipsy. Cows have a complex stomach that breaks down the alcohol in beer, transforming it into nonalcoholic food energy. "By the time the beer reaches the cow's real digestive system, it's a totally different product," says Rochefort. "Cows can absorb many gallons of beer without any increase in the blood-alcohol levels."[21]

It is a wonder that beer isn't used as animal feed more often. According to Barry Robinson, a farm nutritionist with Great Northern Livestock, beer is an excellent source of energy for cows. It also contains vitamins, minerals, amino acids, carbohydrates, and proteins, which may benefit the animal's diet. "A lot of people have joked that the cows look contented, but the beer is just food to them," says Robinson. "It's a good use for a product that used to be poured down the drain."[22]

Both Robinson and Rochefort agree that everyone benefits from the arrangement. The cows get more food energy, the farmer saves money on livestock feed, and Molson has found an innovative use for stale beer. "We aren't making any profit on this project," says Rochefort, "but we're doing it because we are trying to recycle 100 percent of our brewery waste—and because it's the right thing to do."[23]

Beer-Swilling Swine

Molson isn't the only Canadian beer company using its waste products. The pigs of Fen Farms in British Columbia are slurping a little louder these days thanks to a balanced diet of grain and beer. The porkers toss back over 100 gallons of beer a day as part of a project to develop low-cost liquid feed from beer byproducts. The supplier is Labatts brewery of New Westminster, British Columbia. The arrangement helps solve problems for both parties. First, it reduces costs for pig farmers. Second, it solves an expensive waste-disposal problem for the brewery.

Beth Mason runs Fen Farms with her husband, Paul. Beth, who holds a PhD in nutritional science, came up with the idea and approached Labatts to gauge their interest. "A hundred years ago, in Britain, pigs were fed slop," Mason said. "We've gotten away from that to precision-mixing these boring dry ingredients. What we're saying now is that since we have to haul the stuff in from the prairies by the ton, we might as well examine alternatives. And one of those alternatives is going back to a slop system."[24]

The project started as an experiment. Mason was not sure the hogs would like beer, but her worries were soon resolved. The pigs are in hog heaven, banging snouts together to be the first in line for the new feed. Yet there aren't any soused sows. The alcohol is removed from the beer waste during the brewing process.

Fen Farms figures that the beer-feeding program has reduced feed costs by about 15 percent. The brewery is a winner, too, because they've found a way to dispose of beer waste that was being flushed down the drain at a sobering cost.

The Future of the Brewers-Grain Industry

The beer industry has solved its spent grain disposal problem by partnering with the agricultural industry. The problem wasn't solved by more regulations or subsidies for recycling, but by the desire to reduce costs and create new sources of revenue. Nearly all of the spent grains produced by industrial and craft beer makers now end up in the belly of a bovine. Ironically, the future of a successful spent grains market now lies in the hands of Washington, D.C., where lawmakers always seem eager to pass more legislation.

In 1990, Congress amended the Clean Air Act to address smog-related problems. Under the changes, areas with poor air quality are required to add chemicals called "oxygenates" to gasoline to improve combustion and reduce emissions. The act has two programs that require the use of oxygenates, but the more significant of the two is the reformulated gasoline (RFG) program, which took effect in 1995. According to the Congressional Research Service, 82 counties with a combined population of 55 million citizens are required to use reformulated gas. About 30 percent of the gasoline sold in the United States is reformulated.[25]

The law requires that RFG contain at least 2 percent oxygen by weight. Refineries can meet this requirement by adding a number of ethers or alcohols. Two of the most common oxygenates are methyl butyl ether (MTBE) and alcohol made from distilled grains known as ethanol. MTBE is by far the more commonly used oxygenate. In 1999, 87 percent of RFG contained MTBE.[26] Cost is the main reason this particular additive is so widely used.

Over the last few years, however, incidents of drinking water contamination by MTBE—particularly in California—have raised concerns and led to calls for restrictions on its use. By 2003, California will have phased out MTBE entirely. Other states have enacted phase-out legislation as well. And of course Congress has jumped on the MTBE bandwagon with federal legislation intended to eliminate the use, which Congress previously mandated.

At a minimum, the Clean Air Act should be modified to allow for other methods of meeting the performance standard. The act is quite prescriptive in its standards, leaving little room for innovation by the industry.

The act requires the use of oxygenate additives to meet the air quality standards for reformulated gas. Several refiners, however, are experimenting with cleaner-burning fuels that do not require additives. Tosco and Chevron, two firms with large investments in the California gasoline market, have requested changes to the law to allow the sale of reformulated gas that meets the standards, but does not adhere to the oxygenate requirement.

In October 1997, Tosco expressed concern about the growing evidence of the potential for groundwater contamination from MTBE. The company requested that the California Air Resources Board take "decisive action" to "begin to move away from MTBE."[27] Chevron quickly followed with a statement that it "may be possible to make a cleaner burning gasoline without oxygenates, and still reduce emissions to the same extent achieved with current standards."[28]

Assuming that MTBE is ultimately banned, refiners will be forced to find alternative sources of oxygen for reformulating gasoline. An interesting coalition of Midwestern corn farmers and environmentalists are touting ethanol as a viable alternative. Some environmentalists claim it is the silver bullet to both smog and water quality problems. However, a report commissioned by Governor Grey Davis of California after announcing the phase-out of MTBE concluded that if ethanol was substituted for MTBE, there would be "some benefits in terms of water contamination" and "no substantial effects on public-health impacts of air pollution."[29] Furthermore, a study showed the risk of groundwater contamination from both ethanol and MTBE are about the same for the first five years.[30] Finally, as we will see in chapter seven, ethanol may lower carbon monoxide emissions, but it also raises other air pollution problems in its use. Substituting ethanol for MTBE does not appear to be the best solution. Considering the work of Chevron and Tosco, it appears the best policy for reducing air pollution from gasoline may be no mandate at all.

Corn farmers see the ban on MTBE as a boon to depressed commodity markets. The shift to ethanol to fulfill the oxygenate requirement would undoubtedly drive demand for corn and other grains needed to produce ethanol. "We're expecting production to reach 7 million tons by 2004 due to ethanol production," commented Charlie Staff, executive director of the Distillers Grains Technology Council.[31] The Distillers Grains Technology Council may be the only ones happy about the growth. Several of the largest members of the council are Archer Daniels Midland

(ADM) and Cargill, which have the most to gain from federal programs designed to increase ethanol production.

Ethanol plants ferment corn to produce ethanol. Much like the brewing process for beer and alcohol, they produce massive amounts of spent grains. The ethanol plants are looking to the same markets for disposing of their leftovers. The added grain brought on by an ethanol mandate and subsidy will make it harder for existing brewers and distillers to market their spent grains. Miller Brewing Company is watching with great interest, closely monitoring the construction and development of ethanol plants. The amount of spent grain brought to the market by ethanol production would nearly double the quantity currently generated by the beer industry.

The Distillers Grain Technology Council does not think the added grain will cause any problem for beer manufacturers. According to the Council, spent grains are the fastest-growing source of livestock feed in the United States, and only a fraction of livestock producers currently use spent grains. The council believes that with some work, hungry cows, sheep, and hogs could consume all spent grains. "We need to market better to the livestock industry," commented Charlie Staff.[32] Time will only tell if good marketing can overcome the well intended, but misguided, government edict for ethanol.

Saving the Chesapeake

Dealing with problems of waste products by turning them into valuable inputs for agriculture is not unique to the beer brewing industry. On the eastern seaboard, a new way of thinking is aiding the nation's largest and most biologically diverse estuary, the Chesapeake Bay, by turning farmers' fields into waste-consuming machines. Bringing the necessary technology to the problems of the Chesapeake is a company by the name of Sheaffer International. It models its operations on the dictum of company founder Jack Sheaffer, who contends that "Wastes are resources out of place."[34]

In 1983, the governors of Virginia, Maryland, and Pennsylvania joined the mayor of Washington, D.C., and the director of the Environmental Protection Agency (EPA) to create a commission for preserving the Chesapeake Bay. Worried about the Bay's deteriorating health, the group signed an agreement four years later to reduce nutrient runoff in the Chesapeake watershed's main channel by 40 percent. This included 40 percent re-

Table 3.1 Ethanol Production in the United States

State	Number of Plants	Tons of Spent Grain
California	2	16,000
Colorado	1	4,800
Iowa	7	81,600
Idaho	2	19,200
Illinois	5	3,216,000
Indiana	1	272,000
Kansas	4	35,520
Kentucky	1	38,400
Louisiana	1	64,000
Minnesota	15	750,720
Missouri	2	96,000
North Dakota	2	33,600
New England	7	1,404,800
South Dakota	5	268,800
Tennessee	1	144,000
Washington	2	24,640
Wisconsin	2	15,040
Wyoming	1	16,000
Totals	61	6,501,120

Miller Brewing Company 2001[33]

ductions in both phosphorus and nitrogen discharges, most of which came from agricultural runoff.

Another five years passed before the agreement was amended in 1992. The amendments refocused the multi-state compact to shift the goal from cleaning the Chesapeake's main stem to cleaning the Chesapeake's upstream tributaries.[35] Then in June 2000, yet another new Chesapeake agreement was reached.

Titled Chesapeake 2000: A Watershed Partnership, the newest agreement was drafted with the intention of guiding the Chesapeake watershed's restoration through the year 2010.[36] The 2000 agreement includes specific goals for forest management, eliminating wetland losses, increased oyster and crab production, and numerous other environmental proposals. But improved water quality remains the focal point for the latest agreement. In order to achieve that improved quality for the Chesapeake, the agreement reiterates the push for lowering nutrient emissions into the region's rivers and streams. As Chesapeake 2000 states:

> Improving water quality is the most critical element in the overall protection and restoration of the Chesapeake Bay and its tributaries. In 1987, we committed to achieving a 40 percent reduction in controllable nutrient loads to the Bay. In 1992, we committed to tributary-specific reduction strategies to achieve this reduction and agreed to stay at or below these nutrient loads once attained. We have made measurable reductions in pollution loading despite continuing growth and development. Still, we must do more.[37]

Despite a continuing effort to lower nutrient levels in the watershed, the multiple Chesapeake agreements have yet to achieve their stated goals. The 40 percent reductions have not been met in the main part of the Bay or in the upstream tributaries. While the state governments and the Environmental Protection Agency continue to search for the answer, Sheaffer International is stepping forward with solutions.

The North Fork of the Shenandoah River is one of the upstream tributaries that the 1992 amendments target for nutrient reduction. It flows through the north central part of Virginia, almost directly west of Washington, D.C., before joining with the South Fork at Front Royal, Virginia, to become the Shenandoah River. The main channel of the Shenandoah then travels approximately 60 miles before dumping into the Potomac River and, ultimately, the Chesapeake Bay.

Along this path through Virginia, the North Fork passes the poultry processors of Wampler Foods/Pilgrim's Pride and Rocco Quality Foods/Shadybrook, numerous agricultural properties, and the communities of Timberville and Broadway. It was on this stretch of the river that the midwestern company known as Sheaffer International opened a unique treatment plant in August 2000.

Headquartered in Naperville, Illinois, Sheaffer's business centers on a unique reclamation and reuse system for wastewater. The system takes in municipal and industrial sewage for treatment, but unlike conventional wastewater treatment plants, Sheaffer's treatment does not end with the discharge of nutrient-rich wastewater into nearby streams. This is good news for the North Fork of the Shenandoah as well as for those who support the goals of Chesapeake 2000.

Before the Sheaffer system, the North Fork suffered increased nutrient loads from the two food processing plants and the municipalities. With increased nutrients comes an increased risk of eutrophication. Eutrophication is a process whereby plant life grows rapidly, leading to algae blooms or other high concentrations of microscopic organisms developing on the surface. This prevents light from penetrating and limits the oxygen absorption necessary for underwater life. Fish kills and the death of subsurface plants in need of oxygen and light can lead to decomposing organic matter, which in turn leads to more nutrient production, thereby worsening the cycle.

The four emitters along the North Fork employed conventional sewage treatment methods that led to the dumping of over 200,000 pounds of nitrogen and phosphorus into the Chesapeake Bay watershed each year.[38] Sheaffer offered a significant reduction in those nutrient loads, elimination of wastewater treatment odor, and, on top of that, the company promised to do it for less than conventional wastewater treatment.

Dr. John R. "Jack" Sheaffer founded the company bearing his name with the help of a few partners in 1996. Sheaffer first became interested in wastewater reclamation after completing his Ph.D. work at the University of Chicago in 1966. Sheaffer spent a stint with the university's Center for Urban Studies, where he came up with the idea for his reclamation and reuse system. A 1970 letter from President Richard Nixon congratulated Sheaffer on the unique design. Nixon wrote, "I understand that your imagination and dedication have led to the development of a new and promising approach to sewage disposal, and I want to commend you for your pioneering work in this vital field."[39] Two years later, Sheaffer was serving as the science advisor to the Secretary of the Navy. At the same time, Congress was formulating the Clean Water Act. Due to his expertise, Sheaffer was brought in to help write the legislation.

In the late 1970s and early 1980s, Sheaffer helped form an engineering company that created systems to employ the biogases that result from the anaerobic digestion of animal waste for energy creation. The high interest rates of the time and a recession killed the firm.[40] While Sheaffer engineered dozens of wastewater facilities over the years, it was not until the founding of Sheaffer International in 1996 that he had the opportunity to open a business dedicated to his original research on circular wastewater treatment systems.

The Sheaffer system is simple in concept. With only three moving parts, it is easy to maintain. The company's flagship product recycles sewage for agricultural use or landscaping. Wastewater is piped from municipalities and food processors to the Sheaffer system, where it enters a grinder pump. Solids are ground into fine particles and injected into the base of a deep pond called a treatment cell. Anaerobic digestion breaks the organic material down into soluble gases and water. Inorganic material settles to the bottom, where it can be stored for up to thirty years thanks to a slow accumulation rate. On top of this anaerobic zone, a compressor/blower maintains an aerobic zone by injecting air into the cells. The aerobic zone kills pathogens and eliminates odors. After 18 days, the whole process repeats for another twelve days in a second treatment cell. Finally, the wastewater moves to a storage reservoir that is designed to contain it for 120 days with a flow of 1.9 million gallons per day.[41]

This is where the agricultural sector steps forward to do its share for the environment and make an honest dollar in the process. Once treated, the Sheaffer system stores the wastewater until nearby farms are ready to employ it for irrigation purposes. For the North Fork of the Shenandoah, Sheaffer International secured 25-year easements from seven local farmers to spray irrigate roughly 530 acres of corn, soybeans, and hay. The farmers expect corn yields to increase by 30 to 50 bushels per acre thanks to the nutrient-rich water. Soybeans can be expected to yield another 15 to 20 bushels an acre.[42]

To help secure the easements, Sheaffer did not to charge the farmers for the water. One farmer noted that he had planned to put in irrigation anyway, so the free irrigation provided by Sheaffer was a boon. New pivot irrigation rigs, after all, generally cost $500 to $800 per acre to install.[43] It is hard to imagine a farmer who would refuse to take increased yields for no cost other than the easement to irrigate.

Sheaffer signed agreements with farmers based on how close they were to the site, the size of their fields, and soil quality. On other projects, Sheaffer expects to do similar deals with golf courses and other properties demanding regular irrigation. About 200–300 acres of land are needed for a Sheaffer system to reclaim and reuse a design flow of one million gallons per day of wastewater.[44]

By recycling the waste to growing crops, the system avoids discharging into the local tributaries. This works to cut the nutrients in the Shenandoah River and the Chesapeake Bay. The Sheaffer system goes beyond the 40 percent reductions agreed upon by the signers of the Chesapeake

Bay Agreement. With its help, nitrogen discharges by Sheaffer's customers dropped 54 percent and phosphorus loads were cut by 47 percent.[45]

The Sheaffer system is not only an example of waste being taken in by the agricultural sector. It is an example of waste in agriculture going out for a useful purpose. Two of the nutrient emitters that utilize the Sheaffer system on the North Fork of the Shenandoah are food processors, Wampler and Rocco. These poultry producers create substantial wastewater daily. Water saturated with chicken manure includes high levels of nitrogen and phosphorus. What the producers failed to realize before Sheaffer was their waste product has value elsewhere.

This is the genius of the Sheaffer system. It provides benefits for two groups by taking one's waste and turning it into a resource for the other. This helps both sides lower costs while they improve the local environment. Farms lower nutrient discharges by taking in the waste from the food processors, and the food processors provide the farmers with high-valued, prefertilized irrigation water.

The North Fork Opportunity

Dr. Sheaffer recognized that sewage and food processor wastewater includes nitrogen, phosphorus, and potassium in a 35-10-10 ratio. For instance, a typical city of 100,000 people would produce 10 million gallons of wastewater per day and this wastewater would contain about 5,000 pounds of the above nitrogen, phosphorus, and potassium mixture. During reclamation, some of the nitrogen would escape into the atmosphere, leaving the ratio of the nutrients at nearly 10-10-10. Sheaffer realized that this was approximately the same ratio of nutrients in commercial fertilizer. He figured that by not reclaiming the wastewater for irrigation purposes, the conventional discharge process was the equivalent of throwing away 100,000 fifty-pound bags of fertilizer each year. The food processors were creating a potential valuable fertilizer and irrigation product that was going to waste. Worse for them, their industry was suffering bad publicity because of it.

In August 1999, the *Washington Post* ran a three-part series on poultry waste. It pointed to decades of growth by the poultry industry as "the primary source of pollution" in the Chesapeake Bay and other waterways.[46] Those states participating in the Chesapeake Bay compact were in the process of creating legislation to control poultry waste. A year and a half later, the state legislators were still at the process when the Environmental Protection Agency released its proposed regulations for addressing water pollution from Concentrated Animal Feeding Operations, or CAFOs.[47] In its proposal, the EPA estimated that its regulations would result in compliance costs of nearly $1 billion per year and would regulate between 26,000 and 36,000 animal feed operations.

CAFOs grew in numbers across the country as hog, poultry, and cattle operations became concentrated in industrial-type feedlots. Sheaffer estimated that as of 1997, ten firms were responsible for 80 percent of commercial chicken broiler production. As of 1996, four companies were doing 81 percent of beef packaging. And while hog farms decreased from 600,000 to 157,000 over the last decade or so, hog production increased.[48] With an increased concentration of animals, CAFOs suffered from an increased concentration of manure. The poultry producers were feeling the heat as much as anyone, but they had seen the writing on the wall. It was their search for a cost-effective way to dispose of their waste that helped bring the Sheaffer system into the North Fork area in the first place.

In the early 1990s, the Clarke County Environmental Council organized a seminar on the future health of the Shenandoah River. Clarke County is located to the northeast and, more importantly, down river from Rockingham County, where the towns of Broadway and Timberville call home. Two men attended the seminar. In their meeting lay the beginnings of the Sheaffer project on the North Fork. The first man was Bud Nagelvoort, an old friend and business partner of Jack Sheaffer's. Nagelvoort and Sheaffer had worked together at the end of the 1970s on the engineering business that turned animal waste into energy using anaerobic digestion. After that business went belly up in 1981, Nagelvoort became active in Trout Unlimited. He acted as the organization's executive director for several years before retiring in 1992. An active member of the northern Virginia environmental community, Nagelvoort helped create the Clarke County Environmental Council.

The second man at the meeting was a representative from the Virginia Farm Bureau, John Johnson, who would eventually become president of the Virginia Poultry Federation. He was in search of ways to help reduce the nutrient problems, too, but at a reasonable cost to producers. He helped Bud Nagelvoort identify food industry people and municipal interests that wanted to find a solution for lowering nutrients as set forth by the Chesapeake Bay Agreement. Nagelvoort introduced Johnson and the poultry industry to Jack Sheaffer.

After listening to what Sheaffer had to offer, the Poultry Federation worked with the company to identify candidate sites for implementing the Sheaffer system. The proximity of two towns and two food processors with nearby farmlands that could benefit from the irrigation made the North Fork locale superb for a Sheaffer modular reclamation and reuse system. When all was said and done, however, it was the cost that sold the poultry producers and the municipalities on Sheaffer International.

The company offered to finance the construction of the facility in return for charging user fees to the municipalities and food processors that employed it. The town of Timberville estimated the deal with Sheaffer would save it 15 percent on sewage treatment costs.[49] At the same time, because the system is fairly simple, low operations and maintenance costs allow the company to turn a profit. With only three moving parts (the grinder pump, compressor blower, and irrigation pump), the likelihood of something going wrong was low. Thus, most of the maintenance to the system was preventive in nature. Sheaffer estimated that the maintenance and operation costs as well as the construction costs were much less than comparable systems.[50]

The company had a good history to draw upon when it came to maintenance cost. The first Sheaffer system has been operating since 1973, irrigating over 36 million gallons of reclaimed water onto more than 5,000 acres of corn and soybeans each day.[51] In all that time, there has been little trouble.

In addition to offering the company a chance for profit, the structure of owning the facility and charging user fees provided a steady source of income. Sheaffer International believes this arrangement will shield it from business cycles. Ownership also helps the company build its brand name. As one company vice president noted, owning the facility "gives us confidence that our good name will not be besmirched by someone who does not build it properly or operate it properly."[52]

The customers seem to like the user fees, too. Tim Maupin was the manager of environmental affairs for Rocco food processing company when the Sheaffer project came on line. He said, "It's a good deal for us. We don't have to make any capital investments for wastewater improvements and it keeps us focused on what we do best, processing chickens and turkeys."[53]

In addition to water quality, the environment for neighbors of the Sheaffer treatment system is in good shape. As the raw sewage from the municipalities and the poultry producers is never exposed to the air, no odor escapes from the system. The aerated water on top of the treatment cells creates a buffer zone between the natural air outside the system and the foul-smelling sewage being treated. Nothing more than a weak salt-water smell is detectable, and that odor is only noticeable from within a few feet of the system.

Part of the reason that the Sheaffer system is odor-free is that it creates no organic sludge. The obnoxious odor that often comes from typical sewage treatment arises from moving organic sludge through pipes, drying it in open-air pits, and hauling it by truck, train, or barge. But the maceration and decomposition process of the Sheaffer process eliminates sludge. According to a company brochure, "One million gallons of waste-water treated in the Sheaffer Circular System results in approximately 60 pounds of inorganic solids. After treating those same million gallons of wastewater, a conventional treatment system would produce approximately 2,000 pounds (one dry ton) of sludge which must be burned, barged out to sea, chemically processed, or trucked to a landfill where it could leach into the groundwater."[54]

As of April 2001, several other communities are getting in line to benefit from Sheaffer International. The company has at least one other plant planned in Virginia and a dozen projects in some stage of planning elsewhere in the United States, including North Carolina, Ohio, West Virginia, and the Delaware-Maryland peninsula.[55] As long as towns and food-processing companies continue to create waste and farmers continue to demand nutrient-rich irrigation for cheap, Sheaffer plans to profit off of it. And if it helps the environment in the process? Well, the company thinks that's okay, too.

4.

ENVIRONMENT FOR ENVIRONMENT'S SAKE

America's agricultural producers are more than farmers and ranchers raising livestock and crops. They are environmental managers. With more than 55 percent of land in the United States used for farming, no single economic activity compares to agriculture when it comes to impacting the quality of our drinking water, the character of our landscape, and the future of our wildlife.

Many farmers recognize the role they play and already follow practices that protect and enhance the environment. According to a recent US Department of Agriculture study, more than one-third of all farmers have changed the way they plow fields to reduce polluted runoff in nearby streams.[1] Farmers are utilizing a variety of methods to limit fertilizer and pesticide use. In record numbers, farmers are planting trees along riparian banks to protect water quality. Over the last five years, farmers have installed buffer strips of trees and plants along one million miles of stream to intercept agricultural runoff and filter out pollutants.[2] Agrarians are investing their own sweat equity and money into projects that enhance the environment. But what are the reasons for these changes?

Some are implementing change for the dollars from federal handout programs. Those numbers, however, are in decline due not to a lack of interest, but to red tape. Most farmers seeking federal aid to meet public health and environmental challenges are turned away. According to the US Department of Agriculture, three out of four farmers seeking support to change farming practices for the benefit of the environment are shown the door. Similarly, about half of all farmers seeking technical guidance to reduce pollution from agricultural runoff face rejection from Uncle Sam.[3]

The vast majority of farmers and ranchers adopting new, environmentally sound practices are simply doing it for the environment's sake. These individuals are passionate to prove that environmental protection, profitability, and long-term agricultural security are one and the same.

The Milesnick Ranch[4]

Perhaps the two best-kept fly-fishing secrets in all of Big Sky country are Thompson and Ben Hart spring creeks. These two ribbons of water wind their way through a cattle ranch located north of Belgrade, Montana. Below their glass surfaces, the waters team with brown and rainbow trout to excite the most seasoned angler. That was not the case over 30 years ago. Back then, few fish could be found in the creeks. Years of degradation had left them in poor condition.

Fortunately, for fish and anglers alike, the ranch owners Tom and Mary Kay Milesnick found a new way of doing business on their land. The couple is committed to ranching with nature and it is to the Milesnicks' efforts that the streams owe their recovery. For over two decades, Tom and Mary Kay have sought to create a profitable cattle ranch that operates in concert with the environment. "Our first priority is to live by example," shares Tom Milesnick.[5]

The Milesnick ranch began in 1936, when Tom's grandparents purchased its present headquarters near Belgrade. Over the years, the ranch expanded with the purchase of surrounding property, including sections of the two creeks. In addition to its riparian areas, the ranch provides rich, lush pasture for grazing beef cattle.

Grazing practices on the ranch underwent stark changes during the last century. From 1936 to 1970, cattle ran in a single herd over the place, grazing a single pasture on a continual basis. During the winter months, hay and supplement forage guided the herd through harsh conditions. Nearly every ranch in the area followed this same grazing routine. Over time, the cattle trampled the stream banks to create flat shallow stretches that were devoid of fish. The cows also ate most of the vegetation along the banks, which provided the cover essential to protecting fish from predators. The trout population began to decline.

By the mid-1970s, Tom Milesnick recognized problems arising from the ranch's traditional grazing practices. Range conditions on the ranch's summer place near Livingston were deteriorating. The area was never overgrazed, but grass production was stagnant. In response, the Milesnicks developed an intensive rest-rotation grazing program. As part of this practice, cattle are grazed in a small area for short durations. Overgrazing is avoided by moving the cattle once they've eaten the top portion of the plants. Grazing the top portion also stimulates plant growth. Implemented in 1975, the change in management afforded dramatic improvements in range conditions for the Milesnicks and increased beef production by more than 30 percent.[6]

Environmental benefits accrued as well. More elk migrated to the ranch as a result of the improved range conditions. The Milesnicks saw advances in the vegetation around several small creeks. Willows and woody shrubs started to flourish, clear signs of a healthy watershed.

Tom Milesnick worried that environmentalists, concerned about water quality and wildlife habitat, might push for regulations on his cattle. They might try to prevent his livelihood from roaming the stream banks. Rather than fighting them, he decided to prove that cattle, recreationists, and healthy streams could coexist if well managed.[7] First, he needed to improve the health of the riparian area along Thompson and Ben Hart creeks. The dramatic improvements on the summer property led the Milesnicks to adopt a similar grazing program on their place near Belgrade. To do this, Milesnick developed an intensive short-duration grazing program that involved 17 small pastures. The cattle are moved from one pasture to the next. By dividing his pastures into small units, Milesnick is able to give each pasture an extended rest from grazing. Grazing durations ranged from one-half to three and one-half days with up to five grazings per year in any one pasture.

To implement the new grazing program, the Milesnicks had to install new fences. "To better manage stream banks, we fenced out several small riparian pastures using inexpensive singe-strand electric fence, often solar powered," said Tom.[8] This provided a way to limit the riparian areas' exposure to the cattle.

The Milesnicks also changed how the cows used the creeks. They excluded the animals from some of the stream banks while creating gaps in the fence in other places so the cud-chewers could still access the stream for water. In some pastures, cattle were allowed to graze along the stream bank, but were removed from the site before they started eating the grass and plants on the stream bank. Vegetation and grass varieties change as one moves away from the stream and, according to Milesnick, the cattle prefer the grass varieties that grow further away. But once the preferred grass is gone, the cows will move to the forage along the stream. The trick is to move the cattle before they've eaten the favored grass. Moving the cattle from one pasture to the next is easy, though, since they know they are leaving for greener pastures.

Finally, the Milesnicks constructed crossings for the cattle by laying rocks and gravel in a number of small sections of the stream. Concentrating the animals' crossings in one area reduced damage to stream banks. "The cows come from miles to cross at these sections. They seem to like them," claims Milesnick.[9] Just like people, cows prefer to ford where it is easy.

Fishing for Improvements

After improving the riparian areas, restoration efforts were focused on the creeks. The Milesnicks rejuvenated Thompson and Ben Hart by laying rock, removing sediment, planting streamside and aquatic vegetation, and channeling the water to create spawning areas. Results didn't come over-night, but in time, the family created a prime trout habitat in what used to be barren stretches of sand and water. A local fisheries biologist brought in to study the site documented increases in both fish numbers and size.

It did not take long for rumors of the Milesnicks' angling paradise to spread through the fishing community. Fly-fishermen from far and wide started turning up at the ranch in pursuit of the big trout lurking in the depths of Thompson and Ben Hart. For years, the ranch had an open door policy to anglers. Up until 1991, all an angler had to do to gain access was ask permission. But more and more people showed up and according to Tom, "things began to get a little out of hand."[10]

Problems started to occur as the popularity of the creeks grew. The Milesnicks, as well as the hired hands, were finding most of their time absorbed by inquisitive fly-fishermen. That wasn't the half of it. Gates were left open, which made it difficult to maintain the rotation grazing. Beer cans and candy wrappers appeared along the streams, sullying the clean landscape. And some people were using the creeks without asking.

To restore order, a sign-in program was implemented for the ranch in 1991. Simple and straightforward, it consisted of no more than a small box with preprinted permission slips. Anglers signed themselves in and were asked for basic information. One copy of the permission slip went in the box, the other onto the windshield of their car. This made it easy to monitor who had signed in and who was trespassing. Tom could drive around and look for permission slips on car windows. It also made it easier to ascertain who may have littered the property or left open a gate. In fact, requiring the signature alone deterred some bad apples who realized they were no longer anonymous.

The Milesnicks maintained good records, allowing them to see who visited the creeks and how often. One of the rules created with the sign-in program was that anglers limit the number of trips to four a year. Tom and Mary Kay hoped this would curb crowding problems, but enforcing it wasn't as easy as just seeing who had permission slips. As more people started fishing the spring creeks, the Milesnicks had to pay closer attention.

In the end, the sign-in program did little to curb demand or the problems that were arising from angler use. The Milesnicks watched as the number of anglers grew each year. As many as 1,500 fishermen showed up on an annual basis to fish the tiny trout beds. As the numbers grew, so did the abuse of the resource and the privilege of using it. It was an attractive fishing spot not only because of the ongoing restoration efforts, but because access was free. Some anglers were fishing four or five days a week despite the posted request that people limit their visits. Professional fishing guides started bringing clients to the creeks without querying the Milesnicks or offering to compensate them.

The growing popularity began to unravel some of the stream restoration improvements. Stream banks were being trampled. Grass was being crushed. Only this time it was human bipeds causing the damage, not hoofed bovines. With more anglers came more trash. The Milesnicks were picking fast food bags and fishing gear packaging out of fences.

The steady onslaught of anglers meant more headaches for Tom. He was spending increased time monitoring who was on the ranch and whether they had filled out permission slips. All of this was taking time away from the ranching business and cutting into his leisure time. Something had to be done.

From a Free Fishery to a Fee Fishery

During the winter of 1998, the Milesnicks began searching for ways to manage the fishery. Because the two creeks were almost entirely enclosed within his property, Tom decided limiting access was the only option.

Both he and Mary Kay attended several agricultural tourism seminars over that winter in hopes of discovering a solution to their problem. Most of the seminars were geared toward the bed and breakfast industry. They focused on how to attract people to your ranch. They didn't deal with how to turn them away. It seemed the Milesnicks were a special case in the ranching industry, so they sought help from the fishing community.

The main objective was to protect the spring creeks. More than $70,000 was invested in restoration projects and the Milesnicks hated to see their hard work cast away by eager anglers. But closing the ranch to public use wasn't an option either. The family had a tradition of providing access for hunting and fishing and everyone wanted to maintain it. They just needed a way to continue allowing people to enjoy the creeks with neither the congestion nor the monitoring hassles of the sign-in program.

The Milesnicks turned to Bill Bryant, the owner of a local outdoor adventure company. Bryant suggested setting up a fee fishing program similar to the pay fishery businesses operated on the world-renowned spring creeks down the road in Livingston. He arranged a meeting for the Milesnicks with several of those owners. They discussed everything from what to charge to the number of people to allow access each day.

With these ideas in hand, Tom and Mary Kay turned to their good friend Dave Kumlien, a longtime fly-fishing shop owner and respected conservationist. He was instrumental in hammering out a business and conservation plan for the spring creeks. The plan limited use to six anglers per day at a rod fee of $50 per person.

Limiting the number of anglers was a way to prevent the creeks from being over-fished. The number the Milesnicks settled on was an easy number to monitor, which avoided crowding problems and headaches for Tom. With the ranch's size, six people could spread out without bumping into one another.

The rod fee was set to provide modest compensation for managing the pay fishery. "We didn't want to charge too much and limit the locals' opportunities to fish," recalls Tom.[11] The fee allows the ranch to recoup some of its investment in creek restoration projects, while also providing funding for future work planned on Thompson Creek.

The success of the fee fisheries business has been overwhelming. Tom and Mary Kay have established the Milesnick Recreation Company, a business Mary Kay runs from the ranch house. The new business accounts for 7 or 8 percent of the ranch's total revenue, but comprises a whopping 40 percent of net income.

In addition to the financial success, there have been other improvements. Usage of the creeks is down to a manageable 500 to 600 anglers per year. Fishing has been limited to flies and artificial lures, which, a local fisheries biologist indicated, has improved the quality of the fishery. Several portable toilets have been installed and a small fishing hut was constructed to provide relief from the afternoon sun. Trash is no longer a problem, either. According to Mary Kay, there is "no garbage can and there is still no trash on the place."[12]

Educating through Example

The recreation business provides an opportunity for the Milesnicks to improve people's understanding of agriculture. They take every opportunity to tell their story and show how successful ranch management can improve and enhance stream protection. The Milesnicks believe that the best way to help others understand their conservation approach is by inviting people to the ranch.

"Tom Milesnick sees his fishing operation as an opportunity to educate people about ranch life and the willingness of ranchers to be good stewards," says Bud Lilly, a legendary fly-fisherman and vocal conservationist. "On the other hand, he's seeing the dividend that comes with protecting natural assets that the public values."[13]

Throughout the year, the Milesnicks provide tours of the ranch and its creeks to tour groups, media, and school groups. It seems that no one is ever turned away. "I personally know some weeks in the summer that Tom and Mary Kay have hosted one tour per day, each showing how their cattle, wildlife, and fishing benefit each other," said Gene Surber, an MSU extension specialist who has worked with the Milesnicks for over 25 years.[14] The ranch is a source of great pride for the Milesnicks and they enjoy sharing it with people.

"We love to visit not only with other producers, but with non-agriculturalists, and always take the opportunity to explain and show how livestock, wildlife and the environment can exist in harmony with each other," comments Tom.[15] In 2001, he and Mary Kay were recognized for their conservation efforts when they received the Environmental Stewardship award from the Montana Beef Council.

The Milesnicks did not start out with the intent to create a fee fishery. Their goal was to improve the creeks on their ranch. Those improvements, however, attracted hordes of anglers, causing unforeseen problems. Fortunately, this classic tragedy of the commons was overcome by the enforcement of property rights. By closing the fishery to the public and implementing a pay fisheries program, the Milesnicks were able to control the hordes while continuing to provide access to a place they enjoy sharing. Along the way they were also able to recoup some of the costs of rejuvenating their creeks. The money was never that important. In all likelihood, the Milesnicks would have restored the creeks anyway. That's just the way they are. But the revenue from the fishing business has made their job that much easier.[16]

Prairie Restorations

Ron Bowen isn't your typical farmer. He certainly looks the part. He drives a tractor, plows fields, and grows crops. But it is the type of crops that makes him unique. Bowen farms wild prairie grasses and flowers.

President and founder of Prairie Restorations, Inc., Bowen and his company specialize in using native grasses and plants to create prairie landscapes for businesses and home sites. Like so many eco-entrepreneurs, he is pursuing his passion rather than financial reward. He remains passionate about restoring the native grasslands of the Midwest even if it means bringing them to corporate headquarters and new subdivisions. Ron Bowen is in the business of natural landscaping.

Natural landscaping owes its growing popularity to a number of perks. It requires less maintenance, less water, and fewer chemicals, and attracts wildlife by providing habitat and food. Beyond these pragmatics, the visual aesthetic of a natural landscape appeals to many people tired of tightly manicured lawns and blooming flowerbeds.

Prairie Restorations paints stunning landscapes of beauty. At the State Farm headquarters in Woodbury, Minnesota, a vivid tapestry of colors created by Bowen's unique designs recaptures part of the bygone Midwest of Plains Indians, roaming buffalo, and settlers in covered wagons seeking prosperity. While the traditional corporate arrangement relies upon a sea of somber green turf with trees, shrubs, and sterile wood chips, Bowen's work at the State Farm building greets visitors, who react with astonishment at the shift away from vapid verdant and towards vivid vermilion and a splash of violet. Prairie Restorations created the 45-acre plot for the company as a demonstration of its new landscape approach.

"There's no way you'd go into there without saying Wow!" exclaims Bowen. "It's very full of flowers and very beautiful. There are lots of blue and purple tones, a lot of yellow. It's very diverse."[17]

State Farm is one of a growing number of corporate clients for Prairie Restorations. In 1990, IBM hired the company to restore prairie around its computer manufacturing plant in Rochester, Minnesota. Since then, Big Blue has been replacing lawn grass with prairie grasses and wildflowers. Expanding the work to more than 300 acres of its land, IBM now harbors the largest prairie restoration project by a U.S. corporation.[18]

"We had what could be described as a golf-course lawn," said Dick Ulland, an IBM spokesperson.[19] Consequently, the company had to worry about mowing, fertilizing, and other maintenance costs. Not a problem with the new landscape. Prairies do not require as much attention and are much cheaper to maintain. Thanks to the prairie plants, IBM has reduced its ground maintenance costs.

According to Ron Bowen, a company typically spends $1,000 each year to water, fertilize, and mow an acre of lawn. Converting that acre to prairie costs $500 to $4,000, with higher costs reflecting especially fancy landscapes that incorporate lots of prairie flowers. While the cost of establishing the prairie can be high, the annual maintenance costs are a fraction of the traditional lawn's. Typical maintenance for a prairie range costs between $200 and $300 per year.

These cost savings add up. State Farm Insurance spent about $200,000 to plant its 45 acres of prairie. Turf grass would have cost $450,000. That is a savings of a quarter of a million dollars! State Farm estimates it will save another quarter of a million dollars in maintenance during the first ten years of the project.

Bowen got his start in native landscapes while working as a gardener for former Dayton-Hudson[20] CEO Bruce Dayton. According to Bowen, his former boss was an avid horticulturist and knew a great deal about native wildflowers. "Under Bruce, I had quite a bit of free rein. I was able to get a greenhouse and learn how to grow plants and how to manage seed a little bit, and how to do some early restoration work."[21] A 10-year mentorship under Dayton instilled confidence in Bowen, inspiring him to set out on his own. In 1977, he started Prairie Restorations. During its first year, the company had sales of $40,000. In 2001, the company grossed over $1.5 million and employed 15 full-time employees and twenty seasonal workers.

One of the unique aspects of Prairie Restorations that separates it from other landscape companies is its specialty in finding and cultivating rare native plants. Seed production can take a long time. "A lot of times you start with as little as a lunch bag of seed," says Bowen.[22]

Bowen collected his first big bluestem grass seed from an abandoned railroad right-of-way. Since then he has been successful at taking small amounts of seed and parlaying it into a thriving, productive crop. The company planted its first big bluestem field in western Minnesota at the Big Bluestem Farm in 1991. Nine years later they harvested 4,000 pounds of seed. That is enough to plant 400 acres. Seeds from Indian grass, prairie drop, and other prairie flowers are harvested along with the little bluegrass seed. Bowen's mixed prairie grass seed sells for about $15 a pound and it takes 10 to 15 pounds to seed an acre. Potted prairie flowers are a little pricier, ranging from $1 to $4 each.

Corporations are not Prairie Restorations' only clientele. Bowen has designed do-it-yourself kits for homeowners who want to install their own landscapes. He offers a computer program that helps the customers plan a new prairie for their front yard and around the house. Other homeowners, like Eugene Kern, a surgeon from Rochester, Minnesota, hire the company to completely revamp their property.

The customers seem satisfied year round. "I love the flowers," said Kern, another one of Prairie Restorations' satisfied customers. "But [the] prairie has an especially beautiful texture in winter, with the crusted snow all around. The tall grasses stream up as if they were indignant about the weather, and they sway with the wind."[23]

Bowen takes pride in his thriving business, but he seems most proud of the part he plays in restoring some of the flowers and grasses that once covered 250 million acres of prairie stretching from Minnesota to Texas. "Economics sold the projects," Bowen said, "but aesthetics are the greatest reward."[24]

Teller Wildlife Refuge

The nonprofit sector has long been involved in promoting the environment for the environment's sake, but as of late it has been spreading its entrepreneurial wings to help pay for the goals of environmental protection. Groups from the Nature Conservancy to local land trusts are realizing that they have valuable resources and are discovering new ways to tap into the enormous demand that people have for environmental goods. User fees for private wildlife sanctuaries and moneymakers like catered wildlife tours are more common today than ever before. By providing these services and products, environmental groups are moving ecological protection from a philanthropic endeavor to an activity that pays for itself.

Otto Teller was one of the first of the new breed of nonprofit entrepreneurs. He believed that "By taking care of the land and finding new ways to market, we're showing that it's possible to succeed" in environmental protection.[25]

Teller was a strong advocate of land stewardship. He first became hooked on land conservation through his passion as a sportsman. Teller saw an opportunity to put his ideas to work in Montana's Bitterroot Valley, which he first visited in the early 1960s on a fly-fishing trip. He immediately fell in love with the place but became concerned about the number of people moving into the Bitterroot Valley to seek the same solitude that entranced him. More people meant more development.

The growing subdivision of agricultural land worried Otto Teller. To counter this trend, he began buying farms and small tracts throughout the valley during the 1980s and early 1990s and consolidated them into the Teller Wildlife Refuge. He formed the refuge with the notion that conservation practices to protect soil, water, and wildlife habitat would help enhance agricultural value and production as well.

Teller Wildlife Refuge was the first privately owned and managed wildlife refuge in Montana. Today, it consists of 18 properties totaling 1,300 acres along the Bitterroot River. The main part of the refuge includes 880 acres of croplands, uplands, and timber river bottomland. The remaining acreage is scattered along the river within a few miles of the main refuge.

Teller Wildlife Refuge is a study in contrasts. The Bitterroot Valley continues to be one of the fastest-growing areas in the western United States. Unknown to most, the sleepy little refuge provides a quiet corner for human and beast amidst sounds of pounding hammers and screeching saws. The refuge is nearly surrounded by homes, subdivisions, and real estate developments. At the same time, Otto Teller's vision of land stewardship is protecting important habitat and providing a unique opportunity for travelers who are lucky enough to stumble across this special place. The refuge boasts that it offers an "evolving model of how private individuals and small groups can have a significant and positive impact on land use."[26]

Home to a wide variety of wildlife, the refuge offers sanctuary to white-tail deer, red foxes, porcupines, beavers, otters, and muskrats. The refuge also attracts different bird species including osprey, Canada geese, pileated woodpeckers, a variety of waterfowl and raptors, and songbirds that spend their summers in western Montana. Bird watchers flock to the Teller Wildlife Refuge year round.

Expanding the concept of land management to incorporate wildlife, sustainable agriculture, and local community goals requires a number of tools. To achieve this ambitious goal, the site has developed creative educational and research programs that provide visitors with hands-on experiences. Wildlife management, native plant communities, environmental education, stream studies, sustainable agriculture, restoration, and open space protection are all programs of study at the refuge. There is an active volunteer program and plenty of guest facilities to accommodate visitors.

"Our mission promotes an awareness of the place in which we live and respect for the natural web of life," elaborates Diane Boyd, refuge executive director.[27]

Despite the changes that are occurring on its borders, the refuge remains true to its mission of "conservation management, education and research that preserves, protects and enhances the land."[28] A unique aspect of Teller is how it accomplishes this mission through entrepreneurial ventures. The refuge operates a farming business and a thriving lodging and catering business and hosts special events throughout the year.

Farming on a Wildlife Refuge?

Farming is central to the refuge's heritage. Many of the parcels acquired to form the refuge were old homesteads plowed and worked since the mid-1880s. The years of hard use took their toll on the wildlife habitat. Refuge staff remembers the area as "heavily plowed and cowed."[29] But that legacy is changing without abandoning Teller's farming roots. The refuge has undertaken extensive restoration. Large portions of the property are now managed for wildlife alone. Ponds have been constructed, wetlands are reconnected to the river, and native plants and shrubs are reestablished and thriving.

Ten percent of the refuge is farmed using both conventional and organic farming practices. Teller was an organic farmer decades before the current movement caught hold. He began experimenting with organic farming on his Oak Hill Farm in Glen Ellen, California. He purchased the farm with his wife, Eleana Folger, heiress to the coffee fortune of the same name. Teller raised sheep and pears before later switching to flowers and shrubs for the wholesale market. In developing his successful flower business, Teller limited his use of chemical pesticides and fertilizers.

Teller's organic farming interests extended to the Teller refuge, too. Most of the land he acquired for the refuge was placed under conservation easement and one easement condition was that organic farming be used. The refuge leases the land farmed with conventional practices to another farmer, who raises commercial hay and alfalfa, while it takes a more active role in the organic farming operation. Alfalfa and wheat are grown on the lands restricted to organic practices. Since herbicides are restricted under the terms of the conservation easements, alternative methods of controlling noxious weeds are used. For example, food and forage crops for wildlife are planted to help control the spread of noxious weeds.

While small, relative to the acres historically plowed, the refuge's farm operations are central to its "hands-on" research demonstrating the beneficial interactions between farming and habitat protection. "We're trying to show with our farm operations that you can manage for habitat and have a profitable bottom line," explains Boyd.[30]

From a financial perspective, agriculture is a small part of the refuge's income. At best it breaks even. The conventional farming operation provides some revenue, but the organic operations have yet to show any income. Part of the reason for that, explains Boyd, is that the refuge is still learning and experimenting with different organic practices to meet a variety of objectives such as weed control and reestablishing food sources for wildlife. In those cases, maximizing profit is not the main objective, but rather the goal is finding alternatives that lower overall management costs for the refuge.

Teller Wildlife Refuge is still trying to find a fit for farming. In the past, farming occurred on an ad hoc basis with little thought given to its financial and ecological impact. According to Boyd, a number of crops were tried, but with limited financial and environmental success. That is changing, however, under her leadership. Boyd sees a place for farming on the refuge and believes it can be both profitable and an effective wildlife and habitat management tool. As Boyd puts it, "Teller provides a place to set an example for land stewardship and to get people involved and interested."[31] Still, she says, before that can happen, "we need to develop a farm management plan."[32] Even if farming remains a marginal business venture, Boyd finds real value in demonstrating how different farming practices can coexist with wildlife management.

The Teller Education Program

The Teller Wildlife Refuge's active educational program "emphasizes hands-on experiences for kids," says Amy Moneteith, Educational Program Director.[33] The program offers on-site field trips to students in kindergarten through the sixth grade and attracts more than 1,500 students to the small refuge each year. This provides a way for the children to learn and interact with their natural surroundings. Frogs, leaves, birds, and bugs all make for an exciting open-air classroom experience.

A typical day for a student involves visiting four stations positioned around the refuge. Each one of the stations offers a different activity that explains something unique about the refuge. During the winter the refuge staff has to be a little more creative with the station activities. "You've got to keep the kids moving to keep them warm," Moneteith points out, but she adds, "Winter field trips are extremely popular, especially with the kids."[34]

For example, at one of the winter stations the children learn how some animals hibernate during the winter while others migrate to warmer climates. The children are asked questions about different animals on the refuge. Their responses are not written or shouted; rather the children act them out. For instance, they might mimic the motions and behavior of a frog by hopping and jumping around. The game is extremely popular. It is not every day a school child is awarded a good grade for acting like a wild animal.

Special events also play an important part in the refuge's education program, and they're a great way to generate income. The Harvest Festival is the refuge's flagship event for the year. Friends and neighbors from around the community gather at the refuge on the third weekend in September to help bring in the harvest. But this harvest is not like many others that go on in the area. Old-time harvesting practices are demonstrated and used. The Western Montana Antique Power Association demonstrates turn-of-the-century threshing equipment, while a local draft-horse club plows the north 40 with horse and mule teams. Moneteith shares, "The festival is a great chance to show off the different aspects of the refuge."[35]

The Value of Hunting

Hunting was a passion of Otto Teller and when he began to establish his refuge he made sure sporting was included as a part of the wildlife management program. And so Teller Wildlife Refuge is becoming a popular destination for hunting upland birds, waterfowl, and deer. Hunting activities are heavily managed on the refuge, but the end result is a "really nice, low-key, high quality hunting experience."[36] The refuge refuses to charge for hunting despite its popularity. Opportunities are allocated on a first-come first-served basis. When he formed the refuge, Teller believed hunting was becoming a sport of the rich and there were not enough opportunities for the common man. He enshrined his belief by keeping Teller open free of charge, but limiting those opportunities to the twelve people a day who are quickest with the phone. To reserve a hunt, reservations must be set up one week in advance. The first dozen to call secure their chance to bag a bird or a buck. "The phone rings off of the hook during hunting season," says executive director Boyd.[37]

Even with hunting, the refuge understands the importance of incentives. To encourage hunters to volunteer time, the refuge gives its volunteers first crack at the limited number of coveted hunting slots. Refuge volunteers can call eight days in advance, giving them a one-day jump on the rest of the public.

Lodging and Fishing: The Crown Jewel

Aside from its serenity and wealth of wildlife, lodging and fishing opportunities are the crown jewel of the Teller Wildlife Refuge. Several of the original homesteads and cabins scattered about the property have been restored to develop a thriving and lucrative lodging business. For the independent types, a "do it yourself" package is offered at the Ward Cove House and Otto's Fishing Cabin. Both cabins are situated on the Bitterroot River and are well suited to visitors angling for wild Montana trout. The buildings are furnished with unique local character and offer weekly maid service, but visitors have to provide their own meals. Renting at $1,600 to $2,100 a week, it is not uncommon to have to book a cabin one year in advance. For that price, lodgers are assured exclusive use of the river.

Those visitors looking for a little more TLC might want to try the refuge's "spoil you rotten" package at two original homesteads built in the 1860s. These homes offer full maid service, home-cooked meals, and personal attention that rivals even the most exclusive four-star hotel. Both homes are refurbished and outfitted with unique Victorian antiques. Also scattered about the houses are pictures and memorabilia that give a sense of the rich history of the refuge. Living in luxury costs $500 a day for a couple.

The refuge is blessed with great access to the Bitterroot River, one of Montana's renowned fly-fishing streams. A wealth of feisty brown trout makes the fishing a popular draw. Unlike hunting, the refuge has decided to reserve the fishing opportunities for its paying guests. An occasional fisherman off the street is allowed river access when no guests are around, but that rarely occurs during the summer months. Boyd points out that "People pay a lot to stay at the refuge and one of the things they get is use of the river."[38]

Teller Wildlife Refuge is a unique place that is fortunate to have such wonderful resources that people gladly call a year in advance to pay for the experience. The refuge is even more fortunate to have entrepreneurial leadership, which is constantly looking for ways to balance revenue opportunities with environmental stewardship.

Top 10 Reasons for Saving Herring Creek Farms

Herring Creek Farms probably won't make David Letterman's famous top ten list, but the farm surely tops his list of places to stay on Martha's Vineyard. The late-night television host was among a key group of individuals and organizations that came together to save the farm from development.

A $64 million deal including Letterman and Silicon Valley power couple Roger Bamford and Denise Lahey helped save the 215-acre farm in the upscale Massachusetts community from development. The Nature Conservancy, another key player, orchestrated the deal. Under the terms of the transaction, the Nature Conservancy and the Vineyard-based Farming, Agricultural and Resource Management (FARM) Institute, working with Letterman, Bamford, and Lahey agreed to a series of gift and sale transactions to protect the property.

The property was one of the largest unprotected tracts of land on the island and was slated for a 32-lot subdivision. To help finance the deal,

Letterman and the duo from Silicon Valley agreed to purchase two of the limited number of home sites. Both Letterman and the California couple were interested in the success of the deal because they rent homes located on the farm. The Nature Conservancy will sell no more than four additional home sites to generate revenue for paying off the property. For the Conservancy, the deal marks the organization's largest privately funded purchase. "Protecting this piece of property on Martha's Vineyard from large-scale development represents a real victory for The Nature Conservancy and all our partners," said Steven McCormick, president and CEO.[39]

The deal was not only about protecting open space and saving a 300-year-old farm. The property's main attraction is an expanse of native Katama sandplain grassland, a rare habitat that is disappearing all too quickly. "Only 1 percent of the sandplain grasslands ecosystem remains in the world," announced McCormick as he spoke of the deal.[40] Sixty-two acres of the property will be set aside and managed with Katama sandplain grassland in mind. Herring Creek Farm provides a unique opportunity to the Nature Conservancy for restoring lost habitat and uniting the restored ecosystem with existing native habitat.

Transaction Ends Controversy

This transaction ended a decade-long battle between developers and preservationists over the fate of Herring Creek Farm. Though the farm had been in operation for over three centuries, the last time a plow parted its soil was in 1995. For 30 years, Neil and Monte Wallace owned the property. A slowing agricultural economy forced the family to shut down operations and begin considering other options, which included developing the prime piece of real estate. Skirting grasslands and dunes, the farm offers spectacular south-facing views of the ocean beaches. Without the Nature Conservancy stepping in, the farm appeared to be doomed to produce condos instead of cows. Under the new arrangement, residences within the farm are limited to six. Compare that to the 28 new homes included in the last development plan and the 49 new houses proposed initially in 1990. The 1990 plan also included two beach clubs.

Controversy over the property erupted in 1997 when Herring Creek Farm developers challenged zoning restrictions covering the property. The battle landed in the Massachusetts Supreme Judicial Court. The court upheld a three-acre zoning requirement for the farm, citing the threat of pollution to shellfish beds and an endangered coastal sand plain. Environmentalists hailed the decision as a victory, but the farm was still not safe from bulldozers. In November 2001, the Martha's Vineyard Commission, which also had been sued by the developers, gave final approval for a 32-lot subdivision.

The Nature Conservancy recognized that the only way to ensure the fate of the farm was to buy it, so it approached the property owners, brothers Neil and Monte Wallace. According to the organization, the Wallace family was a driving factor in making the deal successful. Neil and Monte donated $18.5 million in cash toward the deal and reduced the property's asking price by an unspecified amount. "The transactions conclude an important chapter in the family's history, the end of its 30-year relationship with Martha's Vineyard and, of course, most importantly with Herring Creek Farm, which the members of the Wallace family have genuinely loved," said Stuart R. Johnson, the family trustee.[41]

The Art of the Deal

This unique transaction would not have occurred without the generous support of several key individuals. A major chunk of funding came via a financial gift from Bamford and Lahey, who made their fortunes with the software giant Oracle. The conservancy retained ownership of 102.2 acres of the farm and restored 62 acres to native sandplain grassland species. After placing conservation restrictions on development, outdoor lighting, and use of pesticides, herbicides, and synthetic fertilizers, the organization sold the remaining 112.8 acres to conservation buyers: Bamford, Lahey, and FARM. The property purchased by the Silicon Valley pair included two lots and an existing farmhouse.

"Of course, we are excited to be involved in protecting such an ecologically important piece of property on Martha's Vineyard," said Bamford.[42]

"The really exciting thing about the Herring Creek project is that people all over the world will be able to see what this partnership accomplished and use it as a pattern for projects at home," added Lahey.[43]

The property acquired by FARM includes a farmhouse, barns, and seven acres of pasture. The organization also will lease 40 acres of the organization's land for a demonstration farm and for education and research. "Sustainable agriculture has historically been an integral part of the community on Martha's Vineyard. This collaboration continues and reinvigorates the island's rich tradition of environmentally compatible farming," said John Curelli, director of FARM.[44] The group is using the farm as a model to demonstrate the options and opportunities for preserving farmland and habitat through active agriculture.

To help finance the group's share of the deal, FARM sold a portion of their acreage to MV Regency Group, a real estate development company owned by David Letterman. David Peters, who manages the MV Regency Group, said on behalf of Letterman, "It is extremely rewarding to see so many people come together to protect this unique piece of property."[45] MV Regency Group plans to sell three lots from the acreage it purchased.

The Herring Creek Farm deal is truly unique. It offers a model for saving valuable habitat even when land is expensive. Thanks to creative financing, some good old-fashioned altruism, and an undying entrepreneurial spirit, the Nature Conservancy managed to pull off an important land transaction in one of the most expensive real estate markets in the country. FARM's director, John Curelli, says the Herring Farm deal "represents a new shift for Martha's Vineyard from development on farmland to preserving it for the environment and the open views it offers."[46]

5.

‍❧

HEREFORDS
AND HABITAT

Only seven to ten inches of annual moisture ever reach the ground in the Paunsaugunt of Utah. Most of the year, the streams of this parched region near the Arizona border carry dust and sand rather than water. Despite these harsh conditions, non-game and trophy game species flourish in and around the desert thanks to the efforts of local ranchers.

Dale Spencer maintains summer range in this part of the world. In 1999, he spent $5,000 making water available not only for his livestock, but also for the large mule deer fast turning the region famous. The Utah rancher cuts back piñon and juniper in order to improve rangeland on his 1,160 acres. He cleans out ponds. To provide forage, he plants alfalfa, clover, bitterbrush, natural grasses, and small burnet. All of these activities benefit the wildlife and all of them are paid for with a check signed by Spencer. According to the philanthropic landowner, "There would be very little wildlife without the livestock operators."[1]

Spencer's generous stewardship is largely due to Utah's Cooperative Wildlife Management Unit program. Through the program, he receives a certain number of tags for big game animals on his property as determined by a survey of the site's animals. He then designates the hunters that the state will sell tags to for use on his land. This allows the Utah stockman to capitalize on the growing market for fee hunting. That has made all the difference for him. "Being able to sell a couple tags helps to pay expenses. Before, it used to just be hunters and wildlife doing damage. The program's changed that all around."[2]

Indeed the program has changed things around. Wildlife is now an asset to be cultivated. The Utah Division of Wildlife gathered statistics finding the average buck deer hunt selling for $3,172. Bull elk brought in $4,125; bull moose, $5,438; and buck antelope, $1,692.[3] With numbers like those, ranchers who ignore wildlife ignore a significant opportunity to help them stay in business.

Sportsmen are willing to pay lofty prices to hunt on properties like Spencer's. Quality wildlife and a high likelihood of a successful hunt draw them in. One portion of the Paunsaugunt sees hunters enjoying better than an 80 percent chance of harvesting a buck mule deer with a rack measuring more than 28 inches across.

With a touch of irony, offering individual animals up to harvest benefits different species as a whole by encouraging landowners to provide habitat. A cycle of better habitat improving wildlife numbers and quality leads to more sportsmen paying landowners for access, which in turn feeds into landowners further improving and maintaining habitat. As agricultural landowners hold the majority of wildlife habitat in the United States, this cycle's creation of ecological agrarians is integral to habitat health.

Agricultural lands are a large portion of the land base in the United States. The 48 contiguous states contain nearly 1.9 billion acres. Of that, approximately 1.1 billion are involved in agricultural activity. Four hundred and fifty million acres have been allocated to cropland uses and 650 million acres are used for pasture to graze livestock.[4]

The habitat provided by private agricultural operations from wheat farms to beef ranches is essential to wildlife survival. Three-quarters of the wildlife in the U.S. live on farm and ranch lands.[5] The federal government estimates that 78 percent of endangered species rely on private land for some or all of their habitat.[6] Michael Bean of Environmental Defense contends, "Without the support of private landowners many species would not have a chance of survival."[7]

The federal government attempts to recognize this through initiatives such as the Conservation Reserve Program (CRP), which pays farmers to let land lie fallow, but it often prevents the agricultural sector from working for wildlife through other measures.[8] The incentives created by taxes and regulatory programs of the federal government can discourage private landowners from maintaining good animal habitat. Actions with the best of intentions often suffer from unintended consequences.[9]

There are, however, private landowners taking the initiative to provide quality habitat for animals on their property despite the poor and contradictory incentives put forth by federal programs. In some cases, the inspiration has come from state programs that help landowners profit from the quality wildlife that their good habitat creates. The entrepreneurs discussed in the next section fall into this category. In other cases, the demand for environmental quality from the marketplace has provided the impetus for habitat provision. This is discussed later with water deals for instream flows, which leave both fish and farmers better off, as the primary example.

Ranching for Wildlife

As the western United States filled up with settlers and fences went up to keep in cattle, a tradition evolved where landowners let neighbors and locals hunt on their property with little more than a handshake in return. In gratitude, some hunters would bring a homemade pie or send the landowner cuts of the game meat upon butchering. Wildlife were plentiful and most farmers and ranchers had no problem with letting the local Joe take a buck from their property.

In time, more people arrived to the West and the advent of better transportation (automobiles, airplanes, and a proliferation of train routes) increased the number of hunters asking for permission. Immigrants and outsiders not familiar with local norms took advantage of the free hunts. A few of these newcomers acted like Joe Six-Pack instead of local Joe. They left open gates, shot up signs, or discarded beer cans on the property. As the bad apples tarnished the reputation of hunters as a whole, the long storied tradition of landowners letting hunters onto their land became a thing of the past. Ranchers and farmers began closing their lands to avoid the costs of generosity.

Hunters were pushed onto public lands and the quality of public lands game stock declined. Over-harvesting and crowding diminish the hunting experience on today's public lands. In desperation, both hunters and state game agencies are returning to the roots of wildlife management. They are looking to private lands for a new future.

Historically, the dichotomy of publicly-owned wildlife relying on private habitat has proven troublesome for state game agencies charged with stewarding state game and non-game species. Wild animals compete with livestock for food and space. They consume the livelihood that farmers sow. With these strikes against them, many in the agricultural sector look at state-owned wildlife as pests, making the notion of creating habitat to help the animals preposterous. This was the case until the mid-1980s when an innovative idea called ranching for wildlife arose in the western United States.

Ranching for wildlife builds on the philosophy of the conservationist Aldo Leopold who wrote, "conservation will ultimately boil down to rewarding the private landowner who conserves the public interest."[10] Jim Tabor of the Washington Department of Fish and Wildlife put it more bluntly, "There needs to be an economic incentive or the habitat won't be there."[11]

The key component defining ranching for wildlife in the eight states is transferable hunting tags. The traditional hunting system in the western U.S. allocated tags to hunters through a state-managed lottery. Landowners could bar hunters from their property or charge them for access, but they could not guarantee hunters the required tag for hunting on their property. With ranching for wildlife, the state provides a landowner with a specific number of tags for the landowner's property. The owner can sell the tags directly to hunters. This provides a link between hunter, owner, and habitat. The hunter can reward the landowner for providing good habitat (and therein healthy species) by paying to hunt again year after year. The landowner can reward hunters who respect the property by providing future hunts.

The number of tags allotted to a ranching for wildlife landowner varies depending on land size, game in the area, and other factors. The landowner then picks those who receive a ranching for wildlife tag to hunt on his property. This is usually done through direct sale, but tags are sometimes given to friends, family, or business associates. The tags are frequently sold as part of a package that includes access to the property and other services from guiding to shelter to transportation. The tags give landowners a valued asset that can be sold for substantial income. This gives them a stake in the game.

In addition to providing the landowners with transferable tags, some programs offer extended hunting seasons. These extended seasons are valued by many landowners. As Ken Mayer of the California Fish and Game Department says, a longer hunting season "allows them to close the ranch down for a week if they so choose."[12] Not only does this make it easier for landowners to integrate the season around their agricultural

Table 5.1 Components of Different Ranching for Wildlife Programs

State Program	Date Established	Extended Seasons	Hunter Permits for Landowner	Minimum Areage	Wildlife Management Plan Required	Habitat Required	Public Access Required
California	1984	Yes	Yes	None	Yes	Yes	No
Colorado	1989	Yes	Yes	12,000	Yes	Yes	Yes
Nevada	1998	Somewhat	Yes	None	No	No	No
New Mexico	See note	No	Yes	None	No	No	Yes
Oklahoma	1992	Somewhat	Yes	1,00	No	No	Yes
Oregon	1995	No	Yes	40	No	No	No
Utah	1994	Yes	Yes	10,000	Yes	No	Yes
Washington	1997	Yes	Yes	5,000	Yes	Yes	Yes

Note: New Mexico's program began in the early 1900s to reward the landowners who helped reintroduce elk into the state. Leal and Grewell 1000: 18.

chores, but as Mayer adds, "when they reopen a week later, it's like the beginning of hunting season."[13]

In return for the benefits provided by the program, landowners make habitat improvements. They often sign management plans for their property, agreeing to make specific changes or maintain certain habitat features in order to enroll in the program. Some states, like New Mexico, eschew the habitat improvement requirement; trusting that tying the number of allocated tags to the number of animals that populate the area will provide sufficient incentive for farmers and ranchers to improve their land. With tags tied to a healthy population, landowners who fail to manage for wildlife numbers receive few tags, just as landowners who fail to manage for quality wildlife herds receive few hunters.

Eight states employ ranching for wildlife to preserve habitat, including the Utah program mentioned earlier. The specifics of the programs are diverse. (See Table 5.1.)[14]

Colorado requires a minimum acreage to enroll in the program. New Mexico does not. Utah requires a wildlife management plan. Oklahoma does not. Washington requires some access for public hunters. California does not. Still, all of these programs encompass the basic component of allocating transferable game tags to landowners in the trust that a stake in the wildlife will improve private landowners' management of habitat and create a slew of ecological agrarians.

The following stories indicate that the trust placed in private landowners by ranching for wildlife programs is well warranted.

California

Bill Burrows and his family have planted dry land alfalfa, cereal grains, sudan grass, perennial grass, and legumes on their property to serve as food plots for elk and deer. They have established water sites for the public's free roaming wildlife. To stimulate waterfowl around reservoirs, they have fenced off access to the reservoirs and created downstream water troughs to break disease cycles. In addition to preserving existing wetlands, the Burrows family built five one-acre ponds, each about six feet deep. Every pond includes an island to provide nesting security for waterfowl. Due to the Burrowses' labors, there are 18 wood duck-boxes and nine mallard tube nests on the property.

The Burrows want to keep their Tehama County ranch in the family. The Golden State's version of ranching for wildlife—the Private Lands

Management (PLM) program—helps them do so by capitalizing on another source of income beyond cattle ranching. Under PLM, hunters, wildlife, and the Burrows all profit from habitat improvements made by the family. Hunters enjoy pursuing and bagging quality game with a higher success rate than hunters on other lands. The wildlife receive improved habitat to survive the increasing urbanization of California. In exchange for the improvements, California's Fish and Game Department rewards the Burrows with 18 deer tags for use on the ranch from the middle of August to the middle of November.[15] The Burrows sell the tags for a healthy sum.

Burrows practices holistic management on his ranch. He says holistic management "is a model, which incorporates three things: quality of life, financial sustainability, and enhanced environments. Quality of life for hunters and guests [is] treating clients like extended family, home cooking, and low numbers to keep the experience on a personal level."[16]

Without PLM, the Burrowses would not have spent $200,000 over the past dozen years on property improvements to provide habitat for a quality hunting experience. The Burrowses, however, are not out to make a killing. "We haven't made any money directly," says Burrows. "Profits go back into land enhancement and wildlife improvements."[17] In addition to spending cash on improving habitat, the family uses the funds to help maintain the ranch. Wildlife and hunters benefit from this reinvestment because it keeps the property in the hands of stewards like the Burrowses.

Down the road from the Burrowses, another ranch stands out in its quest to preserve wildlife habitat. The Corning Land and Cattle Company was worn down when Jeff Weinstein and his partners purchased it. According to Weinstein, "It looked like an alkali salt flat, overgrazed, suffering five years of drought."[18] The property is now a paradise for Tehama County's native species.

Weinstein and his associates burned 700 acres of brush to improve forage. They maintain 15 acres of permanent irrigated pasture to grow a combination of grain, vetch, and forb for wildlife grazing. In return for these and other improvements, the property garners tags for 13 buck deer, 4 antlerless deer as well as 300 upland game bird seals to hunt California quail. Corning has found the business of ranching for wildlife so successful that it no longer ranches cattle. From January to March, Corning leases acreage to others who bring livestock over to graze. The rest of the year, no moo's can be heard at the Corning place. Livestock are entirely removed from the land.

Instead of grazing bovines, Corning offers amenities for deer and quail. There is alfalfa for feed and brush piles for cover. Weinstein and his partners can afford to do this thanks to PLM. Asked whether he would have made habitat improvements without ranching for wildlife, Weinstein echoes Burrowses' sentiments. "I doubt it."[19]

Non-game animals benefit, too. Part of the Burrows agreement with the Fish and Game Department included building four water troughs with access ramps for small species such as snakes and songbirds. The access ramps—made from reinforced-concrete boards one foot wide and six feet long—allow snakes to slither, or birds to hop, down for a drink without falling into the water and drowning. The family also fences livestock out of four reservoirs to allow for bird nesting.

In Mendocino County, the Potter Valley Ranch helped the bluebird population by installing nesting boxes.[20] Another ranch in the PLM program protects peregrine falcon nesting areas. Yet another maintains eight permanent and four portable perch poles to improve Swainson's hawk habitat and built nest boxes for burrowing owls. The Island Mountain Trinity Ranch in Mendocino and Trinity Counties installed blocked culverts, regraded an old road, and filled in gullies to reduce erosion for the improvement of salmon spawning habitat on a creek running through its property. The Tule Land and Cattle Company of Lassen County put flush bars on all of its harvesters to decrease wildlife mortality resulting from agricultural operations.[21] Made from poles with chains dangling down, the bars are hung in front of a farm vehicle so the chains drag through the high grass in front of the machine. They scare birds and small animals out of the way of the farm equipment, preventing them from being crushed or threshed.

The program is successful not only because of the Fish and Game Department and landowners' enthusiasm about the results, but also because the hunters who fund the program are excited. Without hunters buying tags, the economic incentive for landowners to enter ranching for wildlife, and therefore improve habitat, would be lost. Interviews with 55 ranchers both inside and outside the program showed that those participating in the program for its fee hunting opportunities did more to enhance the environment through habitat improvement than those outside the program.[22] California need not worry for now. High success rates, trophy animals, reasonable prices, and access to hunting land key hunter happiness, and California's program provides all four.

Deer hunters on PLM lands boasted a 53-percent chance of bagging a deer compared to only 18 percent statewide.[23] Trophy animals abound on the different ranches in the program. For instance, every blacktail buck deer taken during the 1997 season on the Eden Valley Ranch scored 120+ points on the Safari Club International scoring system.[24] Trophy hunts, however, do come at a steep price. Eden Valley Ranch blacktail hunts retail at $3,500 for four days of guided hunting. Four-day hunts for the rare Tule elk cost upwards of $11,000. While the trophy hunts are more expensive, there are reasonably priced hunts in the program for those willing to do the research.

The Burrows Ranch prides itself on a quality hunt at reasonable prices. For $800, a hunter can get a three-day blacktail deer hunt with all the amenities and a 70-percent success rate. This works out to less than $275 per day and includes meals, lodging, and guided hunt. In contrast, on top of his licensing and guide costs, a hunter could easily spend $40/day on food, $80/day on California hotels and still not come away with a home-cooked meal. A two-day pig hunt on the Burrows' place with the same amenities as the blacktail hunt and a 98 percent success rate sells for $400. Compared to the $858 that the average big game hunter spent in 1996 for travel, lodging, and equipment to hunt, the Burrows' prices seem a bargain.[25] Boone & Crockett scoring animals have been taken on the Burrows spread, but that's not where the family beams with pride.[26] "We don't cater to the trophy market, we cater to quality of life," says Bill Burrows.[27] Considering trophy animals tend to equate with better genetic herd quality, the Burrows can be proud of both attributes.

As the final key to hunter satisfaction, California's PLM opened more private land to hunting opportunity. Sixty ranches were operating in the Golden State's program in 1998. Nearly one-third had been closed to public hunting before the program, while another third still only allow family and friends to hunt. About 20 ranches now allow hunters on their property that had not before the California program.[28]

It is important to stress, as Bill Burrows does, that ranching for wildlife is not game ranching. The animals involved are free roaming, not fenced in. They migrate. They are not selectively crossbred or given drugs like cattle. They are wildlife, not livestock. In stressing the difference between the two, the Burrows Ranch calls the activities on their ranch, "hunting, not shooting."[29]

PLM has excited the agricultural sector about voluntarily improving the environment for all kinds of species. No one is forced to enter the program. It is the financial benefits of eco-entrepreneurial activities such as protecting riparian corridors, planting food plots, and fencing select areas off from cattle that interest the agricultural community.

Colorado

Odd things are afoot at the Twin Peaks Ranch. Nestled along the eastern slope of the Rocky Mountains in southern Colorado's Las Animas County, the 17,000-acre spread isn't run in typical agricultural fashion. Cattle are shut off from grazing in riparian areas. Fields of alfalfa are planted, but hungry heifers never partake of the prime vittles. The ranch trims back oak trees in the region to establish new feed, but not for cows. The Twin Peaks Ranch is managing for wildlife. The operation spends about $15,000 a year on habitat improvements, and that doesn't take into account the time and labor expended or the cost of foregone grazing. The reason for the Twin Peaks' generosity is Colorado's Ranching for Wildlife.[30]

Dave Menagetti is the manager of Twin Peaks Ranch. He says that the owner works hard to steward the property for wildlife. "He does all his own improvements. He bought the ranch for wildlife. He loves to see a lot of wildlife."[31] While it is clear the Twin Peaks' owner has a passion for big game animals, game fowl, and the non-game species, which give his property the wild atmosphere of the Rocky Mountain West, he did not invest in habitat improvements until the property enrolled in Ranching for Wildlife.

Many landowners enter the program because it provides them with tags to sell to in-state and out-of-state hunters for substantial sums. The landowners can charge the hunters for access to their property as well as for amenities like meals and lodging. In Colorado, elk hunting packages on ranching for wildlife lands go for anywhere from $4,000 to $17,000+ for trophy animals. Five-day guided mule deer hunts hover around $3,750 and antelope sell for $3,000.[32]

Twin Peaks, however, hasn't gotten into the tag-selling business yet. It joined Ranching for Wildlife to obtain guaranteed tags for friends and family. Still, Menagetti believes that the ranch will sell tags in the future.

In addition to friends and family, the ranch allows 39 public hunters on the property to take turkey, elk, black bears, and deer. As part of Colorado's guaranteed public access component of the program, these hunters obtain their permits through the public lottery. According to Menagetti, the public is one of the greatest beneficiaries of the program. "I get people saying you guys are lucky to have ranching for wildlife, but I think it's the other way around, the public is lucky. If there wasn't ranching for wildlife, we'd still be hunting the trophy bucks and the public wouldn't be getting on."[33]

Colorado boasts some of ranching for wildlife's most famous patrons. Rick Schroder is known for his role as Newt in the television miniseries *Lonesome Dove* based on the book by Larry McMurtry. The Western bug bit Schroder during filming and he became a real westerner by not only purchasing 15,000 acres of prime elk habitat, but also by keeping that acreage in cattle ranching and hunting. Part of the help for Schroder's efforts has come from Ranching for Wildlife. "I run my ranch as a business. I'm not in it to lose money. If I can afford to only graze 400 cows instead of 800 cows, and make up the income with elk, it enables me to manage the land more gently, because elk are more gentle on the land." Schroder goes on to say, "My ranch couldn't make a profit, protect wildlife habitat and allow for public hunting without Ranching for Wildlife."[34]

Schroder convinced four of his neighbors to go in with him to form the Pinyon Mesa Ranching for Wildlife Association. The association sprawls across 52,000 acres. It has enhanced habitat by restoring aspen stands to the ranches. Streams, ponds, and other water holes dot the landscape thanks to the members of Pinyon Mesa. Reduced cattle numbers and rest-rotation grazing have augmented forage for wildlife.

These improvements are important to the Colorado Division of Wildlife, which runs the program. The improvements boost the local environment's capacity for healthy wild species. In addition to tree trimming, food plots, and cordoning cattle off from the streams and other riparian zones, Dave Menagetti and the Twin Peaks Ranch restrict hunters from traipsing about sensitive parts of the property. Most important to good habitat management says Menagetti is "Overgrazing, overgrazing, overgrazing. We don't overgraze cattle." Ecologically sensitive management is a point of pride for Menagetti who points out, "When I can take cattle out of a pasture and it doesn't look like cattle have even been in there, you know that's good for the wildlife."[35]

Landowners in ranching for wildlife typically earn two and three times the return on the fees for hunting as they did previously on cattle operations. The money to support the program comes from satisfied hunters who enjoy a quality hunt and are willing to pay for it. Data from 1993 to 1997 found that elk hunters in Colorado had a 70 percent success rate on ranching for wildlife lands compared to 38 percent statewide.[36] The success rates for both private- and public-access hunters on ranching for wildlife lands in Colorado were much higher in 2000 than for non–ranching for wildlife hunts as can be seen in Table One. These quality hunts are created by the improvements made by the landowners, which creates the feedback loop mentioned at the beginning of this chapter: better habitat from improvements, more wildlife generation, better hunting, more profits from hunting receipts, and finally more habitat improvements.

Table 5.2 Hunter Success Rates in Colorado for the Year 2000

Success Rates	Deer	Elk
Private Hunters on Ranching for Wildlife Lands	74%	85%
Public Hunters on Ranching for Wildlife Lands	77%	64%
Hunters on non-Ranching for Wildlife Lands	47%	31%

Data from email sent by Jerry Apker, Colorado Division of Wildlife, 18 July 2001

Ranching for wildlife owes its start in Colorado to another well-known persona both in and out of the fee-hunting arena. In 1969, Malcolm Forbes purchased what would become the 180,000-acre Forbes Trinchera Ranch in the eastern section of the cool and dry San Luis Valley of south central Colorado. Forbes hired the father-son team of Errol and Ty Ryland to manage the cattle operation. Things worked as they should for about a decade, but then the ranch began to decline. Cattle prices fell while real estate prices skyrocketed. It seemed the only option was to subdivide Forbes Trinchera and sell it off in bits and pieces as little ranchettes. Fortunately, Forbes had hired two smart cookies for ranch managers. The Rylands proposed saving the property with trophy elk hunts. Following their suggestion, Forbes approached the Colorado game commission in 1986. He offered an intriguing proposal. In return for not subdividing the Trinchera, Forbes asked the state to give him a share of the licenses that it issued for hunting elk in the area. The state agreed with an additional condition that Forbes provide hunter access to the public during the general season.[37]

In the first four years of the program (1986–1989), the ranch made $2.5 million from elk and deer hunting and $210,000 from lodging associated with the hunting.[38] The ranch's prospects were improved. "Without ranching for wildlife," says ranch manager Ty Ryland, "the ranch would likely all be a development. Most of it would have been subdivided and sold off."[39]

The management of Forbes Trinchera now participates in Ranching for Wildlife just like any other ranch. It burns between 2,000 and 4,000 acres to encourage the growth of native grasses each year. It spends substantial sums cutting juniper, piñon, and spruce to encourage grasses and forbs. This provides forage for the animals and fights an erosion problem that was arising on the ranch. The Forbes Trinchera spends close to $1 million annually maintaining its hunting operation and takes in about $1.1 million in revenue from hunting and lodging.[40]

In 2001, the ranch was allocated 85 bull elk and 90 buck deer permits to sell. Its success rate for private hunters averaged, from 1994 to 2000, 85 percent for bull elk and 82 percent for deer.[41] Through the state hunting lottery, public hunters received 75 cow elk tags, ten bull elk tags, ten buck deer tags, and 50 doe tags for use on the property.

Forbes Trinchera has extended efforts to reintroduce other species as well. Work to bring back bighorn sheep paid off when the state authorized the ranch with the only licenses for bighorn sheep hunting in the area.[42] Over the first five years of bighorn sheep hunting, the ranch sold nine permits for $42,000 a piece. In 2001, the public will get the lone permits for big horn sheep on the property as part of a deal giving approximately ten percent of the big horn hunts to the public. Then it will revert back to Forbes Trinchera for several more years before it is the public's turn again.[43] With bighorn hunts selling for so much, the ranch hopes to make back its investment in a short time.

Jarrell Massey has had a similar experience with Ranching for Wildlife. He runs a ranch with his sons, Daryl and Stuart, outside Meeker, Colorado. According to Grewell and Peck, "Deer and elk thrive on the Masseys' ranches, as do beaver, muskrats, game birds, and rabbits. Golden eagles and songbirds, including the mountain bluebird, abound. Hunters are eager to hunt there, even for premium prices, because the land is not crowded and they have a good chance of harvesting either a mature bull elk or a mule deer buck."[44]

The environment flourishes for the Masseys because they work on improving the habitat. Instead of growing wheat as they once did, the Masseys reseeded their land with legumes, grasses, and flowering plants. They created wetland and riparian areas by digging water holes, creating springs, and constructing ponds. Cattle have been kept from prime grazing land so that elk and deer could prosper. The family plants willow saplings and sets fires that reduce sage while encouraging native vegetation.[45] The Masseys' "reputation for improving native habitat for wildlife has also spawned a growing consultant business."[46] In return for their efforts, the Masseys are granted hunting tags. Through their guiding skills, they have turned those tags into successful hunts for paying clients. In 1995, the Masseys received 69 bull elk tags and 24 mule deer buck tags. They filled every one.[47]

One disadvantage seen by hunters when comparing the Colorado program to its sister program in California is that some hunters are priced out of the market. Hunts tend to cater more to the high-paying trophy hunters in the Centennial State because of less competition to fill the different markets. The state of Colorado has a high minimum acreage of 12,000 acres to enroll. This creates some barriers to entry that might keep potential entrants, who could offer lower-priced hunts, out of the program. In addition to raising the price, barriers lower the potential for habitat improvements as landowners who might have participated can not. Nevertheless, the program appears to benefit the environment by appealing to some landowners with a voluntary offer that incorporates a limited property right in a resource with market mechanisms that responsibly allocate the resource.

Washington

David Stevens is a wheat farmer, yet he has taken some acreage out of the golden grain. Located in Grant County, Washington, Stevens replaced the wheat with corn, alfalfa, and sunflowers for the benefit of mule deer. He eliminated livestock grazing, developed springs and water holes, and planted shrubs and trees to provide forage and shelter for wildlife. Electricity used to run sprinklers for the ten forage plots created by Stevens costs him approximately $30,000 a year. Why does he do it? The simple answer according to the Washington farmer, "We like wildlife."[48]

The work done by Stevens and the nearly twenty other landowners in his agricultural cooperative is aided by Washington's Private Lands Wildlife Management Area (PLWMA) program. Through PLWMA, Stevens' property received 50 permits for mule deer bucks in 1998. Of the 50 permits, he gave 20 away (many to landowners in the cooperative), did not use 24, and sold the last six. For those six, he received about $6,000 a permit.[49] Unlike many landowners in ranching for wildlife programs, Stevens admits that he and others in the cooperative would probably make the improvements anyway, but they appreciate that PLWMA's extra revenues offset some of the costs. Stevens actually loses money on the program, because of his high expenses. But there are some benefits. The longer mule deer season under the program—three months compared to just nine days for lands not enrolled—makes it easier to deal with hunters while still running a farm.

The deer seem to appreciate the changes made by Stevens, too. From 1992 to 1998, the deer population on Stevens' Wilson Creek property increased from 1,100 to 1,300. Pheasants and waterfowl, like mallards and geese, have increased on the property thanks to the improvements in wetland and cover areas.

The Washington program has the fewest enrollees of any program. Still in the pilot stages as of 1999, there were only two tree farms participating in addition to Stevens' agricultural cooperative. These properties combined for a total of 200,000 acres benefiting from habitat improvements thanks to ranching for wildlife. Once the program has enough support in Washington to move beyond the pilot stage, the number of participants will probably rise.

❦　❦　❦

In addition to Utah and the other states discussed here, Oklahoma, New Mexico, Nevada, and Oregon have ranching for wildlife in varying degrees. Oregon's program is little more than an expanded landowner preference program started in 1995 to allow landowners the option of giving non-family members one or two tags. On the other side of the coin, New Mexico's program started around the turn of the twentieth century and boasts over 8 million acres and almost 1,500 landowners enrolled.[50] The potential for ranching for wildlife in other western states will increase as those territories without programs witness the success of those with them.

Support is growing. The Greater Yellowstone Coalition has proposed using ranching for wildlife to protect natural habitat along the boundaries of Yellowstone National Park.[51] As Alan Christenson of the Rocky Mountain Elk Foundation pointed out, "If conservation groups do not step forward to work with private landowners, other interests will."[52]

Asked about the best part of the program, Utah rancher Dale Spencer begins to answer, "Selling a couple tags," then laughs a bit, and says with conviction, "no, that isn't really the *best* thing. I like to see wildlife. But the program helps to pay the bills."[53]

The basic concepts behind ranching for wildlife are spreading beyond their humble beginnings. Karl Hess discusses a proposal put forth by the Wildlands Project of Albuquerque and the Gila National Forest in New Mexico.[54] The proposal aims to help wolves. In the proposal, ranchers would voluntarily remove cattle from their federal grazing allotments in exchange for trophy elk licenses that they could sell to hunters. As elk compete with cattle for forage, the Forest Service is often forced to reduce grazing allotments to the detriment of ranchers. But with the new plan, Hess writes, "Everyone wins. The rancher is better off financially, freed from fear of wolf predators and regulatory predation. The hunter has more elk to hunt, for 100 fewer cows can translate into as many as 70 more elk. And the wolf and its supporters gain the most with more abundant prey and a more hospitable ranching environment."[55]

In fact, the ranchers are much better off. Every 100 head of cattle on allotments in and around the Gila National Forest earns about $10,000 worth of net profit. But trading 100 head for four bull elk licenses selling at $4,000 apiece will net the rancher $16,000.

Everyone benefits under the new math of the Rocky Mountain and Pacific West. The whole pie is made bigger, giving everyone a bigger slice. That will continue to be important as undeveloped land in the West continues to grow smaller, leading to outlet malls competing with farms and development competing with the environment. "Conservation on private land is the biggest challenge for the wildlife biology profession," writes Kaush Arha of the Wyoming Game and Fish Department.[56] But the challenge of wildlife preservation does not mean the future must be bleak. For ecological agrarians, like those in ranching for wildlife, challenges are better known as opportunities to make a buck.

Water Markets and Instream Flows

Much like their counterparts on the land, fish species are dependent on agricultural landowners for habitat. Battles between environmentalists and agrarians over water flows often stem from a desire by one side to see fish where the other side sees irrigation water.[57] By these accounts, rancher Rocky Webb makes for an odd environmentalist. With his sweat-stained cowboy hat, faded wranglers, and worn cowboy boots most people would call him anything but a tree-hugger. Yet Webb helped protect steelhead trout in the Buck Hollow Creek, a tributary to Oregon's famed Deschutes River.

Webb was born and raised on his family's cattle ranch in central Oregon along the banks of Buck Hollow Creek. At the turn of the century, his grandfather homesteaded the land, so that he would have a reliable supply of water for irrigation. Water from Buck Hollow Creek helped irrigate a small meadow for growing hay and pasture grass for cattle. But the irrigation created problems for the steelhead trout population.

The stream dries up each year when the ranch withdraws water from the creek. At one time, Buck Hollow Creek teemed with spawning steelhead, but over the years Rocky Webb watched the numbers of fish returning to the creek dwindle. He could remember a healthy population of steelhead and wanted to see the fish thrive again. The rancher had a bottom line to worry about, but he was committed to finding a solution. "Do you wait until the last fish wiggles his way upstream, or do you do something?" asked Webb.[58]

For more than three years, Webb knocked on doors and made countless phone calls to agencies and environmental organizations, searching for a way to help. "I was willing to do whatever I could to restore the fishery, and I was seeking anyone who would listen to what I had to say," he recalls.[59]

Persistence paid off. Eventually Webb found the Oregon Water Trust, a non-profit organization dedicated to protecting fish habitat through market transactions. The Trust and Rocky struck a deal that worked for both the rancher and the fish. He agreed to stop irrigating, thereby protecting the creek from running dry. In exchange, the Trust purchased 78 tons of hay for Webb to feed his cattle. Pleased with the outcome, the central Oregon Rancher-cum-environmentalist commented that he sees the agreement as a way of helping "people realize there are workable solutions out there."[60]

Changing the Way Ag Thinks about Water

Water has played prominently in the development of the West. It fueled the early gold rush. After the miners came settlers who needed water to grow crops. Few people thought about the impact the diversions had on fish and other aquatic wildlife. Their primary concern was settling and cultivating the land.

The miners and settlers developed a system of law for allocating water known as the Prior Appropriation Doctrine. One distinct characteristic of the doctrine is that it allocates water use through a system of property rights. These rights are established through a principle of "first in time, first in right," which means that the first person to put the water to beneficial use is granted a right to continuous use without interference from those who diverted water at a later date.

The development of water rights came at the expense of habitat for fish and wildlife, outdoor recreation, and water quality. The Prior Appropriation Doctrine made it difficult for individuals to establish rights to stream flows. Until the latter part of the twentieth century, water rights were limited to uses that required diversions, primarily irrigation. Furthermore, people had little incentive or need to establish rights for instream uses. That is changing.

In recent years, people have come to recognize the social, economic, and environmental importance of protecting river and stream flows. Like Rocky Webb, people who depend and rely on water for their livelihood are looking for new ways to create incentives to protect instream flows.

Across the nation, agriculture is a major water consumer. Throughout the western states, it accounts for 80 to 95 percent of all water use.[61] With such vast consumption, agriculture is under enormous pressure to reallocate some of its supplies to the environment. But how will this reallocation take place? Some environmentalists feel that agricultural producers should be forced to give up water in favor of the environment. The agricultural community contends that it holds long-recognized property rights to the water, based upon which large investments have been made, so of course farmers and ranchers won't give up water without a fight.

Klamath Falls, Oregon, is ground zero for the battles that may erupt across the West if farmers and ranchers are forced to give up their water.

In years with adequate rainfall, about 150 billion gallons of water flow from the Klamath Lake through a series of canals to water districts that parcel the flow to farms and fields. During 2001, however, the area recorded one of the worst droughts on record. With water in short supply, the Bureau of Reclamation—the federal agency responsible for managing water in the area—had to make some hard decisions. They had to decide between local farmers and two fish protected by the Endangered Species Act. Facing a slew of lawsuits, the agency decided to turn off water to more than 1,400 farmers in order to protect an endangered suckerfish and the threatened coho salmon. This was the first time in nearly a century that the farms would go without water. Outraged by the decision, farmers, on several occasions, cracked open the canals so that water could reach their property.

Local authorities recognize the powder keg that could erupt so they refused to intervene. "They're trying to save their lives," said Klamath Count Sheriff, Tim Evinger, explaining why he chose not to intervene while he watched farmers open canals on July 4.[62]

Water markets offer a way to avoid the animosity that has occurred in Klamath Falls. Through market transactions, farmers are compensated for water that they supply to the environment. Markets also provide farmers and ranchers with an incentive to conserve supplies because water is no longer a "free" right. These developments mean changes for the agriculture sector and offer opportunities for eco-entrepreneurs. The most marked change, however, will be in the amount of water used by the agriculture industry.

Changes have already begun. The value of water has grown so much in the last few years that producers are finding ways to cultivate more land with less water. According to a recent U.S. Geological Survey report, agricultural water use declined for the first time in decades. Over the last ten years, water consumption in nearly every sector has increased, but irrigation use decreased by 3 percent.[63] During that same period, the amount of irrigated farmland increased by nearly 19 percent.[64] These improvements are due to incentives for using better irrigation technologies, driven by water's increasing value.

Agriculture Enabling Water Markets

Recent changes in western water law are creating opportunities for the agricultural sector to enhance the quality and quantity of stream flows through water markets. In one form or another, farmers in every western state can sell or lease their water for environmental needs.[65]

Montana rancher Bob Hanson helped make market transactions a reality in his state. Hanson, a board member of the Montana Farm Bureau Federation, believed that landowners needed an alternative to government regulation to protect fish. To provide that alternative, Hanson coauthored the state's innovative water leasing program, which allows farmers and ranchers to generate income by leasing their water rights to private organizations. "Leasing makes it possible for farmers and ranchers to market their water rights for fish," says Hanson.[66]

Hanson saw the leasing program take effect in 1995 while serving as the Farm Bureau's representative in an unlikely partnership of environmental and agricultural groups. The two sides came together to address water right sales that would benefit fish. Until that time, the groups had been ardent opponents in the state legislature over water issues. The Farm Bureau had opposed any legislation that would result in water being transferred out of agricultural production. The water-leasing program was something both sides could agree on.

The Farm Bureau has good reason to support the leasing program because it is consistent with the group's fundamental commitment to protect private property rights. A basic principle for any property right is that it can be bought or sold. The new law solidifies the idea that water rights are private property rights by allowing them to be traded.

Montana Trout Unlimited and eight landowners on Rock Creek, a small stream in the western part of the state, agreed to the first private lease agreement in Montana. The landowners signed to a 10-year deal with Trout Unlimited to let 1.3 cubic feet per second of water flow past a diversion dam for six months of the year. In return, Trout Unlimited paid the removal cost of an old, unwanted diversion dam. The earthen dam, which created a small pond along the creek, was a liability. "It was failing and needed $50,000 worth of repairs," says Ginny Larson,[67] who with her husband, Jim Larson, helped organize the eight families who leased their water rights to Trout Unlimited. For Trout Unlimited, the lease restored flows in Rock Creek and now provides habitat for native cutthroat trout.

"Jim and I just wanted to see the fish have a healthy environment, and the lease was that opportunity," shares Larson.[68]

It does not take a large amount of water to help out. Agreements like the one crafted by Julie and Larry Williams, farmers from Milton-Free-water, Oregon, are a prime example of how a small amount of water can improve the environment, while providing financial opportunities for agricultural producers. The deal illustrates the willingness and flexibility of private organizations to craft deals that meet landowner needs.

The Williamses struck a 10-year deal with the Oregon Water Trust to let 2.2 cubic feet per second of water flow past their diversion on Couse Creek for six months each year. Now, even in dry years, steelhead trout are able to migrate past the Williamses' diversion. In return, the Oregon Water Trust pays the Williamses for any loss in yield that results from not irrigating their pea and wheat crop. The deal is unique because it allowed the Williamses to use water during years when flow levels were adequate for fish to migrate to the upper reaches of the creek.

For the Williamses, the deal provides money they can bank on. For the Oregon Water Trust, the lease means steelhead trout can return to historic spawning grounds. Market transactions like the one between the Williamses and the Oregon Water Trust are increasingly common.

Fear and Loathing of Environmental Water Markets

Water sales to benefit fish continue to raise some concerns. "This is a good concept but it needs to be approached with caution; that's why we limited it to leases in Montana," says Bob Hanson.[69]

Some farmers and ranchers are wary of water sales, claiming these deals could make it difficult for other irrigators to get the water they need. Water rights contain limitations on how much water can be diverted. But just because a water right allows a landowner to divert a certain amount of water does not mean the landowners will withdraw that much water at all times. In practice, most farmers and ranchers irrigate intermittently, using water only when it is needed for their crops.

When landowners with senior water rights are not irrigating, other right holders are able to divert the remaining water. Some farmers and ranchers fear that their ability to divert unused water will be limited by instream water rights. Under such water rights, the water must remain in the stream. That means when a rancher or farmer sells a water right for instream use, he or she potentially eliminates the opportunity for others to divert unused water when senior right holders are not irrigating.

Rancher Ted Eady from Sisters, Oregon, provides an alternative perspective. "If there's going to be water permanently removed from the land to improve fish habitat, I feel it should come from land that is not the most productive." When this happens, "pressure is taken off other irrigators who ranch or farm more productive land because water needed to restore stream flows won't have to come from their land," says Eady.[70]

Ted Eady raises hay and registered horses near Sisters, Oregon. He struck a deal with the Oregon Water Trust to sell 196 acre-feet of his water right—which dates to 1885—to the Oregon Water Trust for $42,900. Eady had been using the water to irrigate land that he describes as "marginal hay ground."

Squaw Creek, the source for his water right, is habitat for bull trout, a protected species under the federal Endangered Species Act. In addition, the creek flows through the Sisters' city park. Even during years with normal rainfall, the creek would run dry, damaging fish habitat and leaving the park without the benefit of flowing water. Eady's deal with the Oregon Water Trust changed that. He still uses the land, but it now serves as a pasture for horses he purchased with the money from the Trust. "The water is helping bull trout and keeping the city park green, and the money has allowed me to make improvements on the ranch and, more importantly, in my family's life," he says.[71]

Agriculture and Environmental Water Markets

The success of environmental water markets hinges on the agricultural sector reexamining the way water is used and recognizing that new demands are being placed on water. Success also depends on groups, like the Oregon Water Trust, creating financial opportunities to allow farmers and ranchers to be good stewards while still earning a living. Those who value fish and other aquatic species can generate more habitat by making relatively small investments.

The stories described here illustrate how the marketplace is successfully drawing together farmers, ranchers, and environmentalists into mutually beneficial relationships when the incentives are right and the property rights are structured to allow gains from trade. These historic adversaries are also recognizing that the main advantage of markets is their voluntary nature. No one is forced to give up his or her water right, the terms of sales are negotiable, and farmers and ranchers are fully compensated for their rights.

Water markets are not likely to solve all of the West's stream flow problems, nor will markets fit the needs of every agricultural producer. Yet environmental water markets are a move in the right direction.

"People in agriculture aren't against fish, it's just that you can't do anything in this business without irrigation," explains Hanson.[72] Evolving water markets are helping find that balance between environmental stewardship and the bottom line.

6.

THE BOONS OF
BIOTECHNOLOGY

by Gregory Conko

Ever since the publication of Rachel Carson's *Silent Spring*, environmental activists have been warning of a slowly developing but widespread ecological catastrophe stemming from the release of synthetic chemicals into the environment—particularly the use of insecticides, herbicides, and fertilizers for the protection of crop plants. While the use of agricultural chemicals can have negative environmental impacts, predictions of ecological catastrophe have proven to be unfounded. Furthermore, going without those only moderately harmful chemicals means sacrificing tremendous productivity gains and having to bring new, previously undeveloped land into agriculture.

But what if there were another choice? Today, a new crop protection revolution is underway. It is helping farmers combat pests and pathogens more effectively while also reducing humanity's dependence upon agricultural chemicals. Biotechnology[1]—alternatively known as bioengineering, genetic engineering, recombinant DNA engineering, and genetic modification—employs twentieth-century advances in genetics and cell biology to move useful traits from one organism to another. Among other things, biotech enhancements can help plants better protect themselves from insects, weeds, and diseases.

Farmers are enthusiastic about the development of biotech seeds and have made them probably the most quickly adopted agricultural technology in history.[2] By the year 2000, just five years after products of agricultural biotechnology reached the market, farmers around the world planted more than 100 million acres with biotech crops.[3] Just a year later, 26 percent of all corn, 68 percent of all soybeans, and 69 percent of all upland cotton grown in the United States were bioengineered varieties.[4]

It's easy to see why. In the United States alone, the bioengineered varieties of corn, cotton, and soybean, which make up the bulk of adopted biotech crops, have increased yields, reduced agricultural chemicals use, and saved growers time, money, and resources. The increased productivity made possible by these advances allows farmers to grow more food and fiber on less land and with fewer resources. Such productivity gains are essential if we are to outpace the projected increase in global population over the coming decades without putting more land under plough.

In the past, genetic improvement meant combining tens of thousands of genes at random from two or more different parent plants or animals, with little ability to predict what characteristics might result. Today, scientists can identify single genes from one organism, isolate and remove those genes from the surrounding DNA, and then insert them into other organisms, including crop plants or livestock. As the DNA in every living organism is made up of the same basic chemicals—and because DNA works in exactly the same way, whether it's in a bacterium, a plant, or an animal—a gene can be moved from one organism to another and still produce the same trait.[5] Most scientists believe the techniques are actually safer than traditional breeding methods, precisely because breeders know which new genes are being added and what function those genes perform. An analysis published by the U.S. National Research Council in 1989 concluded that

> [Bioengineering] methodology makes it possible to introduce pieces of DNA, consisting of either single or multiple genes, that can be defined in function and even in nucleotide sequence. With classical techniques of gene transfer, a variable number of genes can be transferred, the number depending on the mechanism of transfer; but predicting the precise number or the traits that have been transferred is difficult, and we cannot always predict the [characteristics] that will result. With organisms modified by molecular methods, we are in a better, if not perfect, position to predict the [characteristics].[6]

Dozens of other scientific bodies, including the American Medical Association, the UK's Royal Society, the World Health Organization, and the Institute of Food Technologists have studied modern biotechnology and genetically engineered organisms.[7] They all conclude that the techniques are as safe as conventional breeding methods and possibly even safer.

The introduction of bioengineered crop varieties onto the market has not been without controversy, however. Some critics have suggested that biotechnology could make foods unsafe to eat. But most concerns have revolved around the potential impact of bioengineered crops on the environment. Environmentalists have claimed, for example, that bioengineered crops could harm beneficial insects and other living organisms or become invasive weeds.[8] Much of the public and media attention paid to biotech crops over the past several years has focused on these possible risks. The potential for agricultural applications of biotechnology to improve the environment, however, has been largely ignored except by specialists.

While we cannot claim that genetic changes generated by either biotechnology or conventional breeding pose no risks to the environment, it is important that these risks be put into perspective. On balance, the available evidence suggests that the potential for bioengineered crops and other applications of biotechnology to harm the environment is considerably smaller than the potential environmental benefits. In this chapter I shall focus on the many environmental benefits drawn from agricultural biotechnology.

Although this chapter is not intended to be a comprehensive summation of all the risks and benefits of agricultural biotechnology, I have tried to include discussions of most major themes. The first three sections examine genetic modifications that can directly reduce agricultural chemical use, including insecticides, herbicides, and fertilizers. I primarily discuss bioengineered crops that are currently on the market, though I also include some products that remain in the research pipeline. Next, I examine the broader environmental benefits derived from increased agricultural productivity and some specific advances that will improve yields by allowing crop plants to better resist plant diseases or tolerate extremes of heat, cold, and drought.

Of course, many critics of modern industrial agriculture argue that the choice between biotechnology on the one hand and agricultural chemicals on the other poses a false dichotomy. They argue that organic production methods offer a more environmentally sensitive and sustainable alternative. The end of this chapter documents specific environmental costs imposed by organic agriculture, and concludes that it is, in fact, more destructive to the environment than either conventional agriculture or the biotech alternative.

Pest Resistance and Insecticide Reductions

The use of agricultural chemicals is an environmental compromise. On the one hand, runoff of agricultural chemicals into wetlands, streams, and lakes, as well as seepage of these chemicals into ground water can pose environmental problems. Overuse of chemical pesticides can damage biodiversity in areas adjacent to fields and kill fish or other important aquatic animals, insects, and plants. It can even impact agricultural productivity itself by killing beneficial insects such as bees, other pollinators, and pest-eating insects in and around fields. On the other hand, the failure to use such products can lead to significant yield decline, with its own adverse environmental impacts.

Oerke and his colleagues estimate that, globally, some 42 percent of the world's crops are lost each year to insect pests, weeds, and plant diseases, but that number would climb to 70 percent without pest control. In industrialized countries, such as the United States, some 20 percent of plant productivity is lost to pests, despite the use of pesticides.[9]

One benefit of agricultural biotechnology that has already been demonstrated is its ability to better control insect pests, weeds, and pathogens. Among the most prevalent first generation products of agricultural biotechnology have been crop varieties resistant to certain chewing insects. This trait was added by inserting a gene from the common soil bacterium *Bacillus thuringiensis* (Bt) into the DNA of crop plants. Different strains of the Bt bacterium produce proteins that are toxic to certain insects, but not to animals, including humans.[10] The bacterial proteins occur naturally, and organic farmers and foresters have cultivated Bt spores as a "natural pesticide" for decades. Because the Bt protein is toxic only to insects it was a natural target for investigation. By the end of the 1990s, more than a dozen varieties of corn, cotton, and potato with the Bt protein trait were commercialized in the United States and several other countries around the world.[11]

Consider the success of commercialized Bt corn in protecting plants from the European corn borer (ECB). Depending upon population size in a given year, corn borer caterpillars destroy an estimated 33 million to 300 million bushels of corn annually in the United States alone, causing an estimated $1 to $2 billion in crop damage.[12] The caterpillars are difficult to control even with conventional insecticides, because they actually bore into stalks and ears of corn, where they are not exposed to sprays.

Bt corn has provided farmers with the first truly effective means of controlling ECB infestations. In 1997, when corn borer populations were larger than average, Bt field corn produced approximately 11.7 bushels per acre more, or 8.9 percent, than conventional varieties. During 1998 and 1999, when ECB populations were considerably smaller, the yield advantage was about 4.2 bushels and 3.3 bushels per acre, or 3.1 percent and 2.5 percent, respectively.[13] Thus, while per acre insecticide applications did not decline, Bt field corn has generated more harvested corn with no additional pesticide use.

Another variety of Bt field corn, now awaiting commercialization, is effective against the corn rootworm complex, which costs farmers approximately $200 million in lost production each year.[14] Effective control of the pest is difficult, because corn rootworm has developed resistance to many traditional insecticides, and even to crop rotation strategies.[15] Thus, once it is approved for commercialization, the variety is expected to add to productivity and reduce pesticide use by a substantial amount.

Field trials of a Bt sweet corn variety resistant to the corn earworm and fall armyworm, two economically significant insect pests, showed increased productivity and reduced pesticide applications between 42 and 84 percent.[16] Although sweetcorn varieties have been approved for commercialization, they have not yet been adopted by U.S. farmers.

Engineering Bt into potatoes has also had positive results. One variety, resistant to the Colorado potato beetle, cut pesticide applications nearly in half. Another variety, resistant to both the Colorado potato beetle and aphids, which spread the highly damaging potato leaf role virus, allowed a reduction in pesticide applications from 5.3 treatments per growing season for conventional Russet Burbank potatoes to just 3.2 treatments for the genetically engineered variety.[17] Unfortunately, due to activist pressures, McDonald's and Burger King fast food restaurants told their french-fry suppliers to stop using engineered potato varieties, and the products were removed from the market in 2001.[18]

Bt cotton is perhaps the most remarkable story, generating both substantial reductions in pesticide use and substantial yield increases. Cotton production requires high doses of pesticides in both the U.S. and around the world,[19] so the introduction of Bt cotton varieties resistant to the cotton bollworm, pink bollworm, and tobacco budworm have made a significant contribution to reducing insecticide use. U.S. Department of Agriculture data show that between 1995 and 1999, the total volume of insecticides used to control those three major pests fell by 2.7 million pounds, or roughly 14 percent. An analysis of 1999 harvests of Bt and conventional cotton found an average yield increase of 9 percent with the Bt varieties that year.[20]

Such a large reduction in synthetic insecticide use also saves resources that otherwise would be used in insecticide application. In the year 2000, an estimated 3.4 million pounds of raw materials and 1.4 million pounds of fuel oil were saved in the manufacture and distribution of synthetic insecticides for cotton alone, while 2.16 million pounds of industrial waste were eliminated thanks to the bioengineered cotton varieties. On the user end, cotton farmers were spared 2.4 million gallons of fuel, 93 million gallons of water, and some 41,000 ten-hour days needed to apply pesticide sprays.[21]

In less-developed nations, where pesticides are typically sprayed on crops by hand—and with much less precision—using Bt crops has even greater benefits. In China, for example, 400 to 500 farmers die every year from acute pesticide poisoning. Approximately 50,000 have suffered serious illness attributed to pesticide use since 1987. Since the 1997 introduction of Bt cotton varieties in China, farmers reduced the quantity of pesticides applied to cotton by more than 75 percent compared to conventional varieties and reduced the number of applications from an average of 12 to just 3 or 4 per growing season.[22] As a direct consequence, farmers who planted only conventional cotton varieties reported five times more pesticide poisonings than those who planted only Bt varieties. Small-scale cotton farmers in KwaZulu-Natal province of South Africa enjoyed similar gains. Bt cotton yields were 18 percent higher than conventional cotton during the 1998–1999 growing season, while pesticide use fell by approximately 13 percent. Yields were 60 percent higher the following year, with a 38 percent pesticide reduction.[23]

One final example of Bt's usefulness is in pest-resistant rice varieties, now in development but not yet commercialized. Rice is the major staple food for 2.5 billion people, almost all of whom live in the less-developed regions of the world, where the bulk of twenty-first-century population growth is expected to take place. The International Rice Research Institute (IRRI) in the Philippines estimates that reducing pre-harvest rice loss by just 5 percent worldwide could feed an additional 140 million people.[24]

In 1999, IRRI plant breeders began field testing a Bt rice variety that is resistant to stem borer pests, which cause an estimated 5 to 30 percent yield loss each year. Hence, it was promising to find that field tests in 1999 and 2000 showed the Bt rice yields to be 28.9 percent higher than conventional hybrid rice varieties.[25] This productivity increase was achieved without the use of any synthetic insecticides, which should have a substantial positive impact on surrounding biodiversity.

The Bt Corn Controversy

Unfortunately, Bt crops have been the primary target of many environmentalists, who claim that bioengineered plants could hurt biodiversity. After a brief report in *Nature* showed that pollen from Bt corn can kill monarch butterfly caterpillars, activists began dressing in monarch costumes and protesting against biotechnology. This finding did not surprise scientists, because the corn was engineered specifically to kill caterpillars. Nevertheless, when a second study attempted to simulate field conditions of Bt corn pollen dispersal and found that pollen distribution onto milkweed plants (the only known food source of monarch larvae) in corn fields could be high enough to kill the caterpillars, activists became apoplectic.[26]

However, many experts, like crop scientist Ben Miflin, point out that monarch larvae also die if they are exposed to the Bt bacilli that organic farmers use or to synthetic chemical pesticides. The question is: which method of insect control is least harmful to wildlife?[27] Follow-up studies have concluded that while Bt corn pollen could kill nontarget insects, in actual field conditions the spread of pollen is too small to represent a significant problem for monarchs.[28] Even the U.S. Environmental Protection Agency concluded that all factors "indicate a low probability for adverse effects of Bt corn on Monarch larvae."[29] The gloomy scenario predicted by activists was authoritatively debunked by the publication of six peer-reviewed papers describing two full years worth of intensive field research by 29 scientists who found little or no effect of Bt pollen on monarchs.[30]

Monarch butterflies face much larger threats than Bt corn pollen, including conventional pesticides, the purposeful destruction of milkweed plants, and significant habitat destruction in wintering areas in Mexico and California.[31] Given the potential for reducing the amount of pesticides applied to corn and for boosting productivity, the cultivation of Bt corn and other Bt crop varieties is likely to have substantial net benefits for monarch butterflies and other non-pest organisms.[32] Indeed, scaremongers who continue to fret about the effects of Bt corn pollen on monarch butterflies seem to overlook the fact that monarch populations have actually increased since the 1996 introduction of biotech corn in the United States.[33]

Other concerns have centered on the possibility that Bt protein in soils from roots or crop residues could harm soil biota and beneficial insects. A series of tests conducted by biologists from New York University found that while Bt proteins do persist in soil, they do not appear to have a significant effect on a variety of soil organisms, including earthworms, nematodes, protozoa, bacteria, and fungi.[34] Moreover, fields planted with Bt crops, such as potatoes and cotton, showed larger populations of beneficial insects, because a considerably lower amount of insecticide needed to be sprayed.[35]

This is not to suggest that no environmental harm could ever arise from bioengineered pest-protected plants. To date, though, the evidence depicts an overwhelmingly positive experience with commercialized varieties.

Herbicide Tolerance and Improved Weed Management

Among the most popular traits included in commercial bioengineered crop plants is herbicide tolerance. This allows farmers to apply a specific chemical herbicide spray over fields without damaging the growing crops. The trait can be developed in some plants with conventional breeding methods, but the process is more efficient and effective with biotechnology. Varieties of canola, corn, cotton, flax, rice, and sugar beet, have all been bioengineered to tolerate herbicides; but by far, the most popular herbicide-tolerant crop plant is Monsanto's Roundup Ready soybean. Planted on more than two-thirds of all soybean acres in the United States, this variety is resistant to Monsanto's proprietary glyphosate herbicide, Roundup.[36]

For farmers, the most important benefit of Roundup Ready soybeans seems to be the relative simplicity and efficacy of weed control provided by use of the herbicide-tolerant soybean plants in conjunction with glyphosate.[37] Results regarding the potential for yield increases with herbicide-tolerant soybeans seem to be mixed. Some researchers have documented reduced yields, while others found slightly higher yields. In any case, it is clear that yields per unit of combined inputs are higher, even if yields per acre are not. As Parrott notes, "The savings in cost of production associated with the use of Roundup Ready varieties more than compensate for the lost yield."[38]

Additional questions arise over whether or not adoption of herbicide-tolerant soybeans has reduced the total volume of herbicides used.[39] The exact level of overall herbicide use will remain a contentious matter. One finding upon which all analysts seem to agree, though, is that the use of Roundup Ready soybeans accelerated a shift toward glyphosate, which is generally considered an "environmentally-friendly" chemical, and away from herbicides considered more harmful.[40] This is among the most important, but least appreciated, benefit of glyphosate-tolerant plants.

Studies of herbicide-tolerant cotton and canola, on the other hand, show significant reductions in total herbicide use—approximately 50 percent lower for cotton and 30 percent lower for canola. In addition, fuel cost savings for bioengineered canola due to fewer herbicide applications totaled 40 million gallons in 1997 and 132 million gallons in 2000.[41]

Conservation Tillage and Saving Topsoil

Perhaps an even more important benefit from combining broad-spectrum herbicides and herbicide-tolerant crops is that the system can facilitate the adoption of low-till and no-till farming practices. These cultivation regimens, also known as conservation tillage, are designed to use little mechanical or hand tillage, so the loosening of soil and consequent erosion from wind and water is reduced, by as much as 90 percent.[42] Less tillage results in increased organic matter in the soil, greater efficiency in fertilizer use, and improved soil moisture and fertility. Each of these contributes to higher and more stable yields.

Saving topsoil is necessary for agriculture's long-term sustainability, but eroded soil is also a serious environmental pollutant. Erosion removes more than 12 tons of topsoil per hectare from U.S. cropland annually. When it runs off farm fields, soil can be transported to lakes, ponds, and waterways, where the sediment muddies water, damages aquatic habitat, interferes with navigational and recreational uses, and makes periodic dredging necessary.[43]

Post-planting weed control is a serious challenge for all of agriculture, but it has been specifically identified as a limiting factor in the adoption of conservation tillage practices. The adoption of herbicide-tolerant crops is thus a perfect complement to conservation tillage systems.[44] Farmers can apply one herbicide dose to each field before planting and wait to see where weed pressures develop during the growing season before spraying again. As with pest-resistant crops, fuel consumption and emissions are reduced as the number of herbicide applications falls.[45]

Weeds and Super-Weeds

The primary concern among environmentalists regarding bioengineered herbicide-tolerant crops is that the herbicide tolerance genes could be transferred to wild plants through cross-pollination, creating so-called "super-weeds" that might out-compete other wild plants and destroy biodiversity. There is some chance that genes from biotech varieties could "out-cross" with wild plants, but only in regions where there are wild species related closely enough for ordinary sexual reproduction. Furthermore, this out-crossing is only problematic when the genes in question could enhance the reproductive fitness of the recipient weeds—that is, when it enables the weeds to survive better in the wild.[46]

However, we do not normally spray herbicides on wilderness areas, so the herbicide tolerance trait would not give the wild plant any selective advantage relative to other wild species. Thus, while the transfer of a gene for herbicide tolerance into a wild relative could create a nuisance for farmers, it is unlikely to have any impact on native biodiversity. Even where the herbicide tolerance trait is transferred to weeds, farmers still could control them by switching to a different herbicide, because the trait is highly specific. Indeed, herbicide-tolerant canola plants have been produced with conventional breeding and have been commercially available in North America for more than 20 years. No unmanageable weed problems have been reported as a result of their use,[47] despite several sexually compatible wild relatives often growing near canola fields and canola's proclivity to promiscuous out-crossing.

Fears that the bioengineered plants will themselves become invasive are also misplaced. M. J. Crawley and his colleagues introduced both conventionally bred and bioengineered crop plants into 12 different natural habitats to test whether the biotech varieties persisted longer or were more invasive than their conventional counterparts.[48] The ten-year study included herbicide-tolerant corn, canola, and sugar beet varieties, as well as two insect-resistant potato varieties, in matched pairs with conventionally bred plants. In no case were the bioengineered varieties found to be more invasive or more persistent than the conventional plants, and every bioengineered plant had died off after four years.

Efficient Fertilizers and Soil Nutrient Use

Just as with pesticides and herbicides, the overuse of nitrogen, potassium, and phosphorus fertilizers and the presence of large amounts of animal manures can have negative environmental impacts. Runoff from fertilizers or manures into streams and lakes can cause excessive growth of aquatic plant life, as mentioned in Chapter Three's discussion of Sheaffer International. Stimulating the rapid growth of algae, this nutrient pollution can deplete the availability of absorbed oxygen needed by other organisms. This disrupts aquatic biodiversity by making the water incapable of sustaining large populations of animals, shifting the structure of food chains, impairing fisheries, and sometimes leading to increases in nuisance species populations.[49]

Despite these problems, fertilizers are an important part of food production. "It is fantasy," DeGregori notes, "to suggest that we can grow crops and feed the world's population without some form of crop protection and soil nutrient renewal." In many cases, even newly cleared lands need supplemental nitrogen, potassium, and phosphorus to improve soil quality.[50] Iron and other metals are also important for plant growth. Unlike the nitrogen, potassium, and phosphorus, however, iron deficiency is difficult to remedy with fertilizers.[51] Consequently, research into improving the availability of soil nutrients has long been a goal of plant breeders and genetic engineers.

Some plants, such as those in the legume family (including soybeans, peas, and alfalfa), have developed a symbiotic relationship with *Rhizobium* bacteria that promotes an effective nitrogen "fixation" action. The bacteria convert gaseous nitrogen from the air into ammonia in the soil, which can then be used as fertilizer by the plants. Early research in crop biotechnology often focused on adapting this mechanism for improving plant nutrition in non-leguminous crops. But modifying other plants for such a symbiotic relationship, or to fix nitrogen directly, has proven to be more difficult than initially expected.[52] Research is still continuing in this area, but more promising research efforts have focused on the genetic mechanisms that aid in optimum soil nutrient uptake in certain plants.

In soils that are naturally acidic or alkaline, like those making up over 70 percent of the world's arable land, phosphorus forms compounds with elemental aluminum, iron, calcium, and magnesium. This makes much of the phosphorus unavailable for plant absorption. Many crop plants won't grow to full maturity in alkaline soils unless phosphorous fertilizer is added; nor can they grow to full maturity in acidic soils unless phosphorus fertilizer or lime is added. Nearly 30 million tons of phosphorus is applied every year to farm fields around the world. Even then, as much as 80 percent of what is applied is unavailable to plants because it, too, becomes bound with other elements. With such large amounts of these mineral additives remaining unused by plants, runoff becomes a significant pollution problem.[53]

Scientists at the Center for Research and Advanced Studies in Irapuato, Mexico, have bioengineered tobacco and papaya plants with a gene from the bacterium *Pseudomonas aeruginosa* to secrete citric acid from their roots, which unbinds the phosphorus from other elements and makes it available to the plants. The engineered varieties yield more leaf and fruit than conventional plants when grown in acid soils with no added phosphorus, and they require substantially less phosphorus fertilizer to reach optimal growth.[54] These findings open the possibility that other important crop plants, such as corn, rice, and sorghum—which are commonly grown in acidic soils—could be bioengineered in the same way. Already a similar discovery has resulted in bioengineered rice and corn varieties that grow better in alkaline soils.[55]

Just as with phosphorus, iron is an important plant nutrient that is often present in soils, but not easily absorbed by plants because it is bound with other elements. Rice is particularly susceptible to insufficient iron, so researchers at the University of Tokyo transformed a rice variety with two barley genes to aid in improved iron uptake. The transformed rice secretes mugineic acid from its roots, which frees bound iron in the soil and allows the plants to absorb the nutrient more efficiently. Compared with a non-bioengineered variety, the improved rice showed over 400 percent higher grain yields in alkaline soil.[56]

Toward Less Noxious Manures

Other research indicates that bioengineering can be used to reduce the levels of phosphorus in certain animal wastes. Most livestock, including pigs and poultry, need phosphorus for optimal growth but are unable to utilize the phosphorus in plants. Feed must be supplemented with phosphorus, which leads to high levels of the element in the animals' manures. Researchers at the University of Guelph, in Canada, have bioengineered pigs that can digest plant-based phosphorus, eliminating the need to supplement the animals' feed. The animals, dubbed "Enviro-Pigs," show a reduced fecal phosphorus output of up to 75 percent.[57]

Of equal importance are efforts aimed at making livestock more productive, which can increase the food produced per unit of animal waste generated. Perhaps the best example is the widespread use of a bioengineered version of the bovine growth hormone Somatotropin (bST). Genetically engineered, or recombinant, bST produced in a laboratory is identical to the growth hormone generated naturally by cattle. When administered to lactating cows, the recombinant version of the hormone boosts milk production by 10 to 25 percent for an additional 5 to 15 pounds of milk per day. Although feed consumption must rise to support the additional milk production, there is a 10 to 15 percent increase in feed efficiency. Thus, the use of recombinant bST generates a 23 percent increase in farm profitability and contributes to a substantial reduction in the amount of manure generated per gallon of milk produced.[58]

More Food, Less Land: The Importance of Productivity

The importance of agricultural productivity for ecological stewardship and habitat conservation should be evident. The loss and fragmentation of native habitats caused by agricultural development, along with the conversion of wilderness areas and agricultural lands into residential areas, are widely recognized as among the most serious threats to biodiversity.[59] In a report published by IUCN/The World Conservation Union and Future Harvest, "reducing habitat destruction by increasing agricultural productivity and sustainability" was listed as one of the six most effective ways to preserve wildlife biodiversity.[60]

Over the past 50 years, the world's population has doubled, from three billion to six billion. United Nations demographers expect it to grow by an additional three billion in the next half century.[61] Fortunately for world ecology, farmers have not had to devote substantially more land or labor to agriculture to feed the increased population over the last 50 years. This feat was accomplished by applying scientific knowledge to the development of better plant varieties and animal breeds and the production and better use of herbicides, pesticides, fertilizers, and other agronomic technologies. The result was a dramatic increase in per acre agricultural yields, now known as the "Green Revolution."

The dramatic increase in agricultural productivity generated by the Green Revolution is perhaps the most remarkable environmental success story over the last 50 years. From 1961 to 1993, the earth's population increased 80 percent, but cropland increased only 8 percent, despite increases in per capita food supplies. Increased food demand was met almost totally by increasing per acre yields. Had this not been the case, and agricultural productivity in 1993 remained at the 1961 level, producing the same amount of food would have required increasing the amount of cropland and grazing land by 80 percent or more. In other words, about one-third of the world's land surface—excluding Antarctica—is currently used to grow food. Had the Green Revolution never happened, an additional 27 percent of the world's land area would have had to come into agricultural use.[62] Similarly high yield increases will be necessary in the twenty-first century, though, if the projected population is to be fed with an equally light impact on the environment.

The increases in agricultural productivity over the past few decades have been impressive, but they are not guaranteed to continue. Annual increases in global agricultural productivity have been declining over the last twenty years.[63] Furthermore, the continent of Africa was almost totally excluded from the productivity gains generated during the Green Revolution. Wheat yields in Asia are now four times what they were in 1960 and rice yields have doubled. In sharp contrast, yields of sorghum and millet in sub-Saharan Africa have not increased since the 1960s.[64] Additional productivity improvements in impoverished regions are especially important because, as developing nations advance economically, their citizens are unlikely to be satisfied with subsistence-level diets. Goklany explains that although world population is expected to increase by only half, the larger, wealthier population of the year 2050 can be expected to demand more than twice the amount of food produced today. He estimates that global food demand in the year 2050 will more than double, necessitating an increase of approximately 120 percent above current production levels.[65]

This projected increase in food demand can be supplied in one of two ways: increasing the land area dedicated to agriculture or increasing agricultural productivity. Most of the land that would be available for farming is currently undeveloped habitat. Meanwhile, nearly all of the projected population growth is expected to occur in the poorer, tropical regions of the globe. Often identified as so-called "biodiversity hotspots," these areas house the greatest treasures of biodiversity along with large human populations.[66] Unless steps are taken to ensure that the twentieth-century trend of significant increases in food productivity are carried forward into the twenty-first century, new acreage will need to be brought into production, and wildlife habitat will be displaced.

It will, of course, be possible to achieve additional productivity improvements with conventional breeding techniques. Some scientists believe new breakthroughs are imminent. Nevertheless, most agree that modern bioengineering techniques will be required to achieve the substantial yield increases necessary to fend off large-scale conversion of wilderness area to agriculture. Crossbreeding is a lengthy and imprecise process, and many improvements may not be possible with these older techniques. Bioengineering is much more flexible, precise, and powerful than those earlier methods of genetic manipulation. "We may be able to create the new plant type without biotech," commented Shaobing Peng of the International Rice Research Institute, "but that is where new opportunities will have to come from in the future."[67]

Many scientists are using new discoveries in plant genetics to develop crop varieties that grow faster, producing fruit and seeds in much shorter times. Rice plants, for example, take just over six months to mature in some regions of the world. But researchers from the John Innes Center in the UK are testing a rice variety that matures in less than six months, meaning that two crops can be grown each year instead of one.[68] In addition to increasing yields, delaying the ripening of fruits and vegetables could substantially reduce post-harvest and end-use losses, globally estimated to be as high as 47 percent.[69] Some of the most important genetic improvements—discussed in the next section—will permit crop plants to cope better with severe environmental stresses, including plant diseases and extremes of heat, cold, and drought.

Taking the Stress out of Crop Productivity

As important as pest and weed control and soil nutrients are to crop productivity, controlling the destructive forces of nature do not end there. Plant pathogens, such as viruses, bacteria, and fungi, cause billions of dollars in crop losses worldwide. Scientists have identified several dozen disease resistance genes to a variety of different pathogens, and efforts are underway to transfer the genes into crop varieties. To date, virus-resistant varieties of potato, papaya, squash, and melon have been approved for commercial cultivation, and varieties of citrus fruits, peanut, tomato, and tobacco have all been modified for virus resistance and are awaiting commercialization.[70]

Perhaps the most important examples of a bioengineered virus-resistant crop are two papaya varieties developed by Cornell University researchers. By 1992, the Hawaiian papaya industry had been almost totally destroyed by the papaya ringspot virus, a pathogen that stunts the growth of papaya trees and renders them largely unproductive. Yields of the Cornell papayas are equal to or greater than industry averages, even in regions that had only a few years earlier been devastated by papaya ringspot virus infestation—single-handedly rescuing the Hawaiian papaya industry.[71]

There are many other examples. Field trials in Mexico showed that virus-resistant potatoes boosted yields on small-scale farms by more than 40 percent. In Kenya, yields of virus- and insect- resistant sweet potatoes were nearly 25 percent more productive.[72]

A more difficult challenge to plant breeders has been engineering resistance to a range of bacterial and fungal pathogens. Many fungi can destroy crops, kill trees, and contaminate food or animal feed with deadly toxins, causing some 10,000 different diseases in plants alone. But much progress has been made in recent years in combating the problem.[73] For example, the bacterium *Xanthomonas campestris* causes blackspot disease in peppers and tomatoes. A decade ago, scientists identified genes in some peppers that make them resistant to the disease, but they could not find comparable genes in tomatoes. Using biotechnology, though, University of Florida scientists transferred the genes into tomatoes, making them resistant.[74]

Progress is also being made in controlling the *Fusarium* fungus, which plagues wheat, barley, corn, and other grains crops worldwide. The fungus, which produces the toxic and carcinogenic substance fumonisin, can cause dangerous health problems for both humans and livestock. University of Nebraska researchers have developed bioengineered spring wheat and barley varieties that carry antifungal protein genes resistant to *Fusarium*.[75] Additional research into controlling fungal and bacterial infestations is well under way with bioengineered varieties of potato, tomato, cucumber, and tobacco.

Researchers have also discovered that effective methods of pest control, such as Bt crop varieties, can keep in check viruses and fungi that are spread by insects, including *Fusarium*. Researchers at Iowa State University and the USDA found that by reducing insect damage, Bt corn substantially reduces the incidence of *Fusarium* and fumonisin—in some cases by as much as 75 percent. Since grain contaminated with fusarium must be either discarded or diluted with uncontaminated crops, preventing such fungal infestations can reduce post-harvest losses considerably and add to overall productivity gains.[76]

Watering Plants with a Dash of Salt

Extremes in temperature, periods of drought, and impure water are also significant factors that limit the productivity of crop plants. Advanced genetics and bioengineering are helping agronomists understand the biochemical processes that allow certain plants to thrive in specific environments. These advances will allow breeders to develop crop varieties that thrive despite environmental stresses.

For example, water is essential for growing healthy plants, but about one-third of humanity lives under conditions of relative water scarcity. Making matters worse, more than 70 percent of all fresh water used globally is for farmland irrigation, but more than half of that never makes it to crops because of evaporation and leakage.[77] Thus, the development of crop varieties able to grow in soils with low moisture or to tolerate temporary drought conditions could have substantial environmental benefits by permitting current farmed land to remain productive longer, by boosting yields on those lands, and by freeing up water resources for other uses.

Researchers in Brazil have identified a gene in tobacco responsible for reducing dehydration during periods of drought, and they have bioengineered tobacco plants to overexpress, or strengthen, that genetic response. The transformed plants grow substantially better than non-bioengineered plants during periods of reduced water.[78] Other researchers have identified plant genes that will help crops better survive bouts with extreme heat and cold.[79] Even the improved pest and pathogen resistance traits discussed above will make water use more efficient, since much of the loss to insects and pathogens occurs after the plants are fully grown, and thus after most of the water required to grow the plant will have already been used.[80]

In one of the most promising research advancements, scientists in Japan bioengineered the wild mustard plant *Arabidopsis thaliana*—which is closely related to several common crop species—to overexpress one of its own dehydration response genes. This allows the transformed plants to tolerate several different environmental stresses, including drought, freezing, and salt loading.[81] Such engineering should prove beneficial, because many soils can become salty when water containing mineral salts evaporates from the upper layers of the soil. Over time, excess salinization can depress crop yields, and in extreme cases, high salinity can force abandonment of the land.

About one-third of the irrigated land worldwide is so affected by salinization that it is unsuitable for growing crops, and that amount is increasing by some 200,000 hectares each year.[82] Thus, the development of crop plants that will grow in saline soils could contribute significantly to meeting the growing food demand of the twenty-first century. Early research into saline tolerance led some researchers to believe that the trait could be bred into plants from wild varieties. While some success has been achieved in producing salt-tolerant varieties of tomato and barley through hybridization with wild relatives, most researchers now believe that further gains will only be achieved with advanced biotechnologies.[83] Fortunately, several other recent advances are already leading to the development of salt-tolerant crop varieties.

Scientists at the University of Toronto and the University of California at Davis have engineered *Arabidopsis* for increased salt tolerance and transferred that gene into tomato plants. These transformed plants are so tolerant to salt that they not only grow in salty soil, they can also be irrigated with brackish water with only a modest negative effect on plant growth. The same research team has engineered canola that grows when irrigated with highly saline water. Eventually, the scientists believe, it should be possible to transform every single important crop variety, because most have the same or a similar gene. Furthermore, because the plants take up large amounts of salt into their leaves and stems, it may be possible to use such plants to desalinate salty soils over time.[84]

Additional research on producing salt-tolerant crop plants is under way in India, where researchers are working to develop rice varieties, and in China, where scientists are working to develop eggplant and pepper varieties that could be irrigated with seawater.[85]

The Organic Alternative: Environmental Friend or Foe?

On balance, the data indicate that biotechnology is already contributing to a decline in the use of agricultural chemicals and an increase in productivity. The vast potential of crop varieties and livestock breeds now in the research pipeline suggests that additional improvements are achievable in the near future. Many critics of biotechnology argue, however, that the choice between bioengineered crop varieties and greater agricultural chemical use is a false dichotomy. Organic and other "natural" farming advocates believe that intensive agriculture, which relies upon heavy use of synthetic and other "off-farm" inputs, devastates soil health, makes for unhealthy food of poor quality and taste, and has a serious detrimental impact on the surrounding environment.[86]

Claims that organic farming is better for the environment, however, are difficult to substantiate, because organic practices merely trade some environmental threats for others. For example, organic farms do not generate the same sorts of synthetic chemical runoff as modern, industrialized farms; but organic farms do still need to control pests, weeds, and pathogens. They also need to replace soil nutrients drawn off by growing plants. Judged by the standards of those who criticize modern agricultural practices, the techniques that organic farmers employ to accomplish these tasks are far from eco-friendly.

While organic farmers do not use synthetic pesticides, they do use chemicals to control insects and plant diseases—including such potentially dangerous chemicals as copper sulfate, rotenone, pyrethrum, ryania, and sabadilla.[87] These "organic" pesticides are derived from minerals or plants and lightly processed and thus are considered to be natural for the purposes of organic agriculture. Yet, ounce for ounce, most are at least as toxic or carcinogenic as many of the newest synthetic chemical pesticides.

Copper sulfate, for example, is toxic to humans, highly toxic to fish and earthworms, and is designated a hazardous substance under the Federal Water Pollution Control Act. It is known to cause liver diseases and has killed vineyard workers. Moreover, copper and sulfur residues remain inert in the soil and continue to be toxic and inhibit plant growth for years. Rotenone is toxic to humans and other animals, and it is highly toxic to fish. Pyrethrum has been classified as a "likely human carcinogen" by a U.S. Environmental Protection Agency scientific panel, and it is highly toxic to fish, posing a potential runoff problem. It also has to be used at much higher doses than the synthetic pyrethroids employed in conventional insect control.[88]

With only a few exceptions, these organic pesticides control insects and plant diseases far less effectively than synthetic chemicals, so they must be used in larger doses. All in all, organic farmers use an average of 100 times more of these natural pesticides per acre than conventional farmers use synthetic pesticides.[89]

In addition, because organic farmers must control weeds by using frequent mechanical tillage—or sacrifice yields—organic agriculture contributes to topsoil erosion and disturbs worms and other soil invertebrates.[90] Compared with modern conservation tillage practices, organic weed control is significantly more damaging to an ecosystem.

Instead of soluble nitrogen, potassium, and phosphorus fertilizers, organic farmers rely on animal manure and so-called "green manures," such as legume nitrogen fixation or organic plant matter, to restore soil nutrients. This raises two issues. First, plowing legume crops and animal wastes into the soil leads to nitrate leaching into groundwater and streams at rates similar to conventional agricultural practices.[91] Second, the chemical properties of the soluble mineral fertilizers prohibited in organic farming are identical to those that are released in uncontrolled quantities by the mineralization of organic matter.[92]

Even more important than the problem with organic fertilizers is the land use problem in actually growing organic crops and raising organic livestock. Even proponents acknowledge that organic farming generates per acre yields that are at least 5 to 10 percent lower than conventional crop production and as much as 30 to 40 percent lower for certain types of crops, such as potatoes, wheat, and rye. Organic livestock productivity is approximately 10 to 20 percent lower than for conventional husbandry.[93] Even these figures can be misleading, because soil nutrient replacement on organic farms requires lands to be fallowed with nitrogen-fixing plants, such as clover or alfalfa, for two or three years in every five or six. Conventional farming does not need to fallow fields. Hence, conventional farms can match the yields of organic farms with only 50 to 70 percent of the land.[94]

Conclusion

No single publication, let alone a single book chapter, could ever thoroughly tally the many benefits and potential risks of agricultural biotechnology. But since it is primarily the risks that tend to be communicated to the public, this chapter is intended to put those potential problems into perspective. Only by understanding the benefits of bioengineered products as well as the risks can one adequately judge the value of the technology. The examples described above show that, far from being an unmitigated environmental hazard, agricultural biotechnology offers the opportunity to provide a more secure global food supply with fewer detrimental impacts on the environment.

This chapter only begins to tell the story of the environmental benefits that could be obtained from agricultural applications of biotechnology. Many other products, such as biodegradable plastics and nonpetroleum-based synthetic fibers, can be produced with bioengineered plants. Tree species grown for pulping into wood fibers and paper can be engineered to require less processing. It is even possible to use bioengineered plants to help clean up hazardous wastes.[95]

Although the complexity of biological systems means that some of these promised benefits of biotechnology are many years away, the biggest threats that consumers awaiting the bioengineering revolution face at present are restrictive policies stemming from fears that the technology poses unique and dangerous threats to human health or the environment. The bottom line is that fear and overregulation are slowing progress in agricultural biotechnology and inflating the costs of research and development. Few would disagree that innovators should be cautious, as all new technologies have both risks and benefits. Appropriate regulatory approaches involve weighing the risks and benefits of moving into the future against the risks and benefits of forgoing the new technology. They do not point to hypothetical risks and say no.

7.

THE ROAD TO ECOLOGICAL FAMINE

We began this volume by looking back at agriculture's past. We end it with two chapters looking at the present and peering into the future to see where agriculture is headed in its evolution with the environment. During the rise of modern environmentalism in the 1960s, agriculture often escaped heavy criticism. DDT and other pesticides were the focus of Rachel Carson's famous book, but in general, most of the environmental concerns of the early movement revolved around the industrial sectors more visible to urbanites. Pollution from energy producers, automobiles, steel mills, and chemical plants received the bad publicity. They were the primary targets of legislation such as the Clean Water Act, the Clean Air Act, and the Superfund programs. The biggest problem agriculture faced in the early days was higher prices for pesticides, as several inexpensive pesticides were banned.

But in recent years, agriculture has taken hard knocks from the environmental movement. The poultry industry is blamed for polluting many of the nation's waterways. In 1999, a three-part series in *The Washington Post* called poultry the "primary source of pollution reaching key portions of the Chesapeake and coastal bays of Maryland, Virginia and Delaware."[1] The Environmental Protection Agency lays 60 percent of the pollution in rivers and streams and 45 percent of lake pollution at the feet of agricultural sources.[2] The agency cited this damage when proposing regulations at the end of 2000 to regulate concentrated animal feeding operations, or CAFOs, and the new Total Maximum Daily Load regulations, which will apply to all waters.[3] Between fertilizers and raising livestock, agriculture is the leading source of nutrient loading in the nation's waterways, providing about 36 percent of nitrogen in U.S. waterways.[4]

Overuse of pesticides and herbicides can harm biodiversity. As we saw in the last chapter, even the natural chemicals used by organic farming—considered benign by many—threaten wildlife and human health. Tillage practices erode essential topsoil. The dams that provide much of the irrigation water for farms and ranches are now blamed for ruining fish habitats. Summer 2001 witnessed a contentious fight over irrigation water in Klamath Valley, Oregon, as farmers vied with endangered suckerfish and coho salmon for the rights to water.[5] In North Carolina, hog farms are the topic of debate. Both the smell, and the pollution into waterways, upset environmentalists, neighboring landowners, and downstream fishers. In a 1999 *Newsweek* editorial, an attorney for the Natural Resources Defense Council wrote that hogs in North Carolina outnumber the citizenry and produce more fecal waste than all of the people in California.[6]

Today, there are fewer farms and ranches than five years ago, ten years ago, or even 50 years ago, but the size of operations that remain in business is expanding. This creates increased awareness of the above problems, as a few concentrated polluters are more visible than many dispersed, but smaller, polluters. As Mark Jenner of the Farm Bureau writes, "Commercial livestock facilities have grown in size and concentration, and so has the size of their waste streams. This has created both legitimate and perceived fears about livestock facilities. Environmental concerns revolve around nutrients, pathogens and odors."[7]

Our narrative has focused on agriculture's evolution led by agrarian entrepreneurs who are facing these challenges from the environmental movement and trying to carve out a niche that will help their bottom line and contribute to the health of their land. These industrious souls do not work in a vacuum, however. They are affected by the institutions around them.[8] How they manage their businesses and the resources they choose to value are products of the rules and market forces by which they must abide. The effect of institutions can be helpful when they clarify the "rules of the game" or allocate property rights in resources as we saw in the case of wildlife and water in chapter five, but they can also be destructive to the entrepreneurial process. They can slow the evolutionary process and hurt the environment by giving landowners the wrong incentives for managing their land.

In agriculture, as in many industries, the institutions that cause the problems are often the result of well-meaning people. The road to Hell is paved with good intentions. Good intentions can create poor incentives that might not lead to a destiny of fire, brimstone, and demons bearing pitchforks, but can cause serious harm to agrarian businesses and the environment.

To provide a broad overview of what we mean, let us discuss a variety of institutions affecting the agricultural sector and the environment. For their relation to agriculture and the environment, we consider three different programs: the Federal Estate Tax, the 1973 Endangered Species Act, and agricultural subsidies, including ethanol and the Conservation Reserve Program. After discussing how the institutional arrangements of these programs have paved a road harmful to the environment, the chapter that follows will offer suggestions for institutions that can help the environment and agriculture.

The Federal Estate Tax

The Federal Estate Tax explains how institutions (the law or the rules of the game) can create perverse incentives negatively impacting farms, ranches, and ecosystems. Congress instituted estate taxes and repealed them three times over the course of the eighteenth and nineteenth centuries. Every time, the purpose was to garner funds for military buildup during times of war. When the estate tax, also known as the death tax, was reintroduced for the fourth time in 1916 to provide funds for World War I, it managed to stick around for the long haul.[9] Over the years, its value to the federal coffers declined as the rich found ways to avoid paying the tax through complex estate planning. By 1998, the Joint Economic Committee of Congress stated that "the tax produces no benefits that would justify the large social and economic costs" and "the estate tax raises very little, if any, net revenue for the federal government."[10]

While the value of the estate tax declined to the government, its level of harm rose for property holders. Private tree farmers and traditional agricultural producers have few liquid assets but significant property holdings, so they bear more and more of the estate tax burden with increasing difficulty. In 2000, the heirs to an estate owed taxes when they inherited $675,000 worth of assets. Due, in cash, nine months after death, the minimum estate tax rate equaled 37 percent, nearly the rate of the highest income tax bracket. One million dollars is a substantial inheritance, but for agricultural owners, the inheritance generally occurs in the form of pasturelands, farm equipment, and other farm assets—not liquid cash. Dirt rich, but dollar poor; farmers and ranchers must sell off equipment and land essential to running their business. This tax hurts not only the viability of agriculture, but natural ecosystems, too.

In a study cited by the Joint Economic Committee, the Keystone Center in Washington, D.C., found:

> Federal estate tax requirements are a major obstacle for private landowners whose land stewardship has been sensitive to environmental value and who would like to be able to pass on their land to their heirs without destroying that value. The imposition of federal estate taxes often forces large parcels of environmentally valuable land to be broken up into smaller, less environmentally valuable parcels. Some of the best remaining habitat for endangered species is put at risk in this manner.[11]

The situation of Martha Clark offers apt illustration. In 1981, Clark's father-in-law died, leaving her the family farm in Maryland near Washington, D.C. The estate came with a hefty death tax and Clark was forced to sell 250 acres of the farm to subdividers to pay the amount due. With family farms struggling to stay profitable, fragmenting the land to pay the taxes can sink the operation. As Martha Clark put it, "If you had a warehouse and you sold off one corner of it to keep the rest, would your warehouse still work?"

Farms like the Clarks' provide the majority of open spaces and forage areas for deer, antelope, foxes, wild turkeys, and other creatures. They provide migration corridors unimpeded by highways or housing developments. When they are sold off or subdivided into strips, wildlife suffers. The property becomes fragmented. New fences, buildings, and roads are erected, and animals find migration corridors cut off. Foraging grounds end up under concrete or grassy lawns. According to Michael Bean of Environmental Defense, "Federal estate tax requirements are destroying some of the largest and most important endangered species habitats in private ownership."[12] Endangered species are hit hard by the tax. As was noted in chapter five, the GAO reported in 1994 that 78 percent of endangered species used private land for some or all of their habitat.[13] An example of this occurs in the Sunshine State, where only 30 to 50 Florida panthers are believed to still exist. Florida biologist Dennis Hammond intimates the death tax may be leading to the extinction of the rare panther because private lands that serve as habitat are being fragmented by the estate tax.[14]

Even the Speaker of the House for the United States Congress is not exempt from the estate tax and its destructive qualities. During a floor debate in 2000, Speaker Dennis Hastert recalled how the tax prevented him from taking over his Illinois family farm when his father-in-law passed away. "Every tractor, every combine, every roll of fence, and every head of cattle was sold off so we could pay the estate tax. I might have been a good farmer, but I didn't have that choice," recounted Hastert.[15]

To this point, no comprehensive study has been done on the tax's effects on agriculture. One study, however, has examined the tree farm industry. Researchers from Mississippi State University and the U.S. Forest Service estimate that about 350,000 acres of nonindustrial forestlands are destroyed each year to pay off federal estate taxes. This is equivalent to half the acreage of a Yosemite National Park, annually. Lands that were once wildlife habitat become housing developments, shopping malls, and parking lots as owners who are short on cash sell off property to pay their federal bill. Private timber owners sell 1.4 million acres of forestland every year to pay the estate tax and one quarter of that acreage ends up employed in a developed capacity.[16] In fact, the estate tax may even be biased toward urbanization, as it is most likely to develop in areas where land prices are high or rising. Land prices tend to be highest in areas undergoing urban growth, so the heirs can not afford to keep farming or growing trees due to the massive tax bill.

Such private forestlands often provide the only cover for animal wildlife. Without the trees, numerous bird species have no place to nest. Even if land is not sold to pay the tax, habitat can end up altered as a result. The Mississippi State/Forest Service study estimates that each year 2.6 million acres of forestland are harvested to pay the tax.[17]

As part of the $1.2 trillion dollar tax cut that President George W. Bush signed in 2001, the death tax will be phased out over the course of a decade. In 2010, there will be no federal estate tax. But farmers who want to pass on their farm to heirs who will keep farming better die that year, because in 2011 the tax plays the part of the phoenix, rising from the ashes and coming back into force unless new legislation repeals it.[18] As an institutional constraint, the estate tax creates the wrong incentives for farmers and ranchers to help the environment and to stay in business.

The Endangered Species Act

The Endangered Species Act of 1973 (ESA)[19] exemplifies how good intentions can have dire consequences for the environment and agricultural landowners. The purpose of the act was to save at-risk species from extinction. With such a high-minded goal, it passed with virtually no opposition. The bill sailed through the House of Representatives with a vote of 355–4 on December 20.[20] In the Senate, a voice vote passed the legislation; an earlier vote on the Senate version of the bill had passed 92–0.[21] Before December ended, President Richard M. Nixon had signed the bill, declaring that the ESA "provides the Federal Government with the needed authority to protect an irreplaceable part of our national heritage—threatened wildlife."[22]

Among the powers and duties bestowed upon the federal government, Dean Lueck finds five of particular significance:

> First, the secretary of the interior was required to establish a list of species, subspecies, and/or isolated populations that were endangered or threatened. . . . Second, the act made it unlawful to take any endangered or threatened species whether on private or public land (section 9). More important, *take* was broadly defined to mean "harass, harm, pursue, hunt, shoot, wound, kill, trap, capture or collect" an endangered species (section 3). Third, the act required federal agencies to review all their actions and to make certain they did not jeopardize any listed species or modify critical habitat (section 7). . . . Fourth, the act allowed courts to award court costs for plaintiffs (section 11). Fifth, the ESA dramatically extended federal authority into the management of resident wildlife traditionally held by states.[23]

For our discussion of unintended consequences, Lueck's second power and duty is the most important new aspect created by the ESA. Initially, many thought the ESA limited a *take* to the direct harming of an animal.[24] In 1975, however, the Secretary of the Interior defined "harm" to include, "significant environmental modification or degradation which has such effects."[25] The significance of the secretary's definition was not regarded with much concern at the time, though at least one article recognized a possible impact on the rights of private property owners to manage their land.[26] It was not until the 1978 *Tennessee Valley Authority v. Hill* case that the true power of sections 3 and 9 of the ESA were recognized. The case forced federal agencies to engage in no action that would "result in the destruction or modification of habitat of such (endangered) species."[27] In essence, defining harm, the case affected section 9 of the Act so that habitat alteration on private lands would qualify as a take, too. After that case, the Fish and Wildlife Service began to assert ever-increasing powers to regulate the use of private lands when endangered species were found in residence. After years of conflict, habitat alteration on any land was reaffirmed as a take in 1995 by the Supreme Court's decision in *Babbitt v. Sweet Home Chapter of Communities for a Greater Oregon.* The court agreed that "habitat modification or degradation where it actually kills or injures wildlife" qualified as harm under the ESA.[28]

The evolution of federal government power to regulate private landowners' management of their land via the ESA morphed innocent animals into the enemy of private landowners across the country. For agricultural landowners, an ESA listing became more deadly to their livelihood than a hailstorm in June. The federal government could restrict the use of property, thereby lowering its value without giving the landowner anything in return.

The ESA allowed the federal government to tell landowners what they could and could not do on their property. Landowners were encouraged by the institutions that evolved under the ESA to manage against animals already on the brink of extinction. Rumors of private landowners implementing the three-S method of endangered species management (shoot, shovel, and shut up) were disregarded as anecdotes and hearsay at first. But in 1994, wildlife legal scholar Michael Bean of Environmental Defense admitted in a training session for government employees that private landowners were "actively managing their land so as to avoid potential endangered species problems."[29] Discussing the regulations set up under the ESA, Bean went on to say, "They're trying to avoid those problems by avoiding having endangered species on their property."[30]

Two years later, Environmental Defense put forth a paper that asked why the United States was losing the battle against extinction. The authors described landowners' fear of increasing land use restrictions from endangered species populations. They wrote, "In its most extreme manifestation, this fear has prompted some landowners to destroy unoccupied habitat of endangered species before the animals could find it."[31] The authors had identified preemptive habitat destruction.

The most famous instance of preemptive habitat destruction occurred in the early 1990s. North Carolina landowner Ben Cone became embroiled in a battle with the red-cockaded woodpecker. Cone had nothing against the bird. In fact, he had managed his timberlands to help wildlife. He would clear-cut a 50-acre block every five or ten years to simulate the effect of a small fire and did controlled burns to clear back the under story.[32] While not managing for woodpeckers in particular, his goal of helping deer and quail species helped the woodpecker, too. A few birds began to breed on his land. Unfortunately for both Cone and the red-cockaded woodpecker, the bird was on the endangered species list.

When Cone decided to sell some timber in 1991, he hired a biologist to do a survey of the birds. Twenty-nine birds in 12 colonies were found on his property.[33] The U.S. Fish and Wildlife Service informed Cone that he had to draw a half-mile radius around each of the colonies and could not cut timber there. That regulation cost Cone $2 million in property value.[34] Over 1,500 acres of his land had come under the powers of the U.S. Fish and Wildlife Service and he could not do a thing about it.[35] He could, however, do something about the other prime woodpecker habitat that had not yet been infested with the birds. He began to cut down trees. "I cannot afford to let those woodpeckers take over the rest of the property," Cone said. "I'm going to start massive clear-cutting. I'm going to a 40-year rotation instead of a 75- to 80-year rotation."[36] By switching to a quicker rotation, Cone would destroy the old-growth pine habitat that woodpeckers need to survive. It would keep the birds off his property and the rest of his land free from regulation. It had the complete reverse effect of what those who created the ESA intended. It hurt the red-cockaded woodpecker's chances of recovery.

Cone was not alone in feeling the burden of the Endangered Species Act. Anderson and Leal describe the travails of Dayton Hyde, a rancher in eastern Oregon, and his bout with the Act.[37] Hyde was the kind of eco-entrepreneur hailed in this book. He worked hard to build a better ecosystem on his 6,000-acre spread. A lake with a three-and-a-half mile shoreline, teeming with rainbow trout and thousands of migrating birds, owed its creation to the landowner. He restored wetlands on 25 percent of his pastureland; land previously drained for hay fields. After spending $200,000 on improvements for wildlife, he started an organization with the sole goal of improving private habitat.

In the process, Hyde's ranch attracted the standard bearer of all endangered species, the bald eagle. A rare species of algae also grew on his property, one of only six isolated spots in the whole world the algae called home. The beauty and the bounty created by Hyde also brought a federal bounty of a different kind down on his property. By housing the endangered eagles and algae, Hyde's property fell under the jurisdiction of the ESA. Hyde hoped to recoup some of his investment by subdividing his land and selling select areas to people for retirement homes, but between the Act and Oregon's land use laws, Hyde's plan was prohibited. As Hyde said, "federal regulators had the authority to shut me down anytime, without compensating me."[38] He could afford some lost value to his property, but he knew that was not the case for everybody. "Ranchers don't have a lot to play with when it comes to making a profit. Take some of their assets or prevent them from using some of their assets, and they can easily go broke," he added.[39]

Cone and Hyde's stories are not enough to establish evidence of systematic behaviors created by the institutions formed from the ESA. Their problems could have been the exceptions to the rule. But a study by economists at Montana State University and North Carolina State University found systematic and credible evidence that the ESA harms listed species by encouraging preemptive habitat destruction.

Professors Dean Lueck and Jeffrey Michael conducted research in North Carolina, home of the red-cockaded woodpecker. They surveyed the owners of more than 400 forest plots to determine the average age of their timber at harvest. They collected data on the number of woodpecker colonies within a 25-mile radius of each landowner's property. By comparing timber harvest age with the closeness of the colonies, they found that landowners with the most colonies nearby cut their timber nine years earlier than normal and industrial timber operations cut 42 years earlier in the most heavily populated areas than in the least populated areas.[40] Because the woodpecker's preferred habitat is old-growth pine, landowners were trying to discourage the endangered birds from making homes in their trees and, thereby, avoid federal regulation of their land. (See Figure 7.1).

Lueck and Michael's study may not be evidence of shoot, shovel, and shut up, but it does indicate that preemptive habitat destruction is occurring. This is the see, saw, and sell-early phenomenon: see the birds nearby; saw down your trees; sell your timber early.

Figure 7.1. Predicted Harvest Age of Industry Timber by Number of Red Cockaded Woodpecker Colonies within 25-Mile Radius

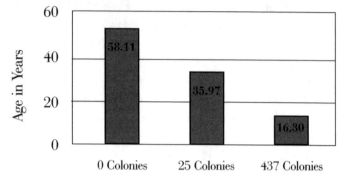

Lueck and Michael 1999.[41]

This may be why the ESA reads like an obituary column instead of a recovery ward. At the end of June 2001, there were 1,244 species listed as threatened or endangered under the Act (737 plants and 507 animals) and many more proposed.[42] In 1997, *Environment International* published analysis by researchers at the National Wilderness Institute who examined the 27 species delisted since the ESA's inception. Seven species were removed due to extinction. Fourteen never warranted listing in the first place. Of the remaining six, the brown pelican and the arctic peregrine falcon owe removal to the banning of DDT, the California gray whale's population was improving before listing, and three kangaroos were saved by Australian policies.[43]

As the stories of Ben Cone and Dayton Hyde illustrate, the ESA has trained timber owners, agriculturalists, and other landowners to fear endangered species. The unintended consequences of its well-meaning institutional arrangement create incentives for private landholders to engage in behavior exactly contrary to the Act's goals.[44] Institutional constraints such as those of the Endangered Species Act work against the efforts of eco-entrepreneurs and discourage, or even punish, members of the agricultural sector from evolving into environmental helpers.

Agricultural Subsidies

Over the years, subsidies may have done the most damage to our natural environment. In agriculture, they encourage the use of marginal land, they make environmentally damaging forms of production more appealing than other forms of production, and they discourage investment in higher yields on plots of already-cultivated acreage by encouraging the use of additional land.[45] Lands that otherwise would have been unprofitable for farming have been cultivated because subsidies made certain property more valuable in agriculture than out of it. Gardner writes, "High guaranteed prices, received by farmers and administered through price-support programs, have increased the values of agricultural land. Land has been cultivated at the extensive margin that would have remained in rangeland and forests, especially in the southern region and in the semiarid and arid regions of the Great Plains and Rocky Mountains."[46] Environmental harms don't end with land conversion.

In Florida, the sugar subsidy has damaged the Everglades by lowering the water table to irrigate marginal sugar fields. As Paul Roberts explains, "This sweet protectionist deal not only adds a nickel profit to every pound of sugar produced by large U.S. cane farmers but has abetted the Everglades' decline by encouraging marginal swamplands that could not be profitably planted otherwise."[47] Clay Landry agrees:

> Federal protectionist policies for domestic sugar growers have further exacerbated environmental problems in the Everglades. Tariffs on imported sugar and federal price supports for domestic sugar encouraged expansion of sugarcane production in the region. The federal government diverted water and drained more lands to make room for more sugarcane. As a result, the Everglades now receives less than one-third of its historic water flow.[48]

The Green Scissors coalition reports that sugar production destroys three to five acres of the Everglades every day.[49] Over half a million Everglades' acres have been converted from swampland to sugar fields.[50] Much, if not most, of this sugar production occurs in the region only because of the federal subsidies and the billions spent on government water control projects. The minimum price for sugar, which is guaranteed by the U.S. Department of Agriculture, reached nearly three times the world price during 1999.[51] The U.S. General Accounting Office estimates that the federal sugar program cost consumers of domestic sweeteners $1.5 billion in 1996 and $1.9 billion in 1998.[52] From 1980 to 1988, the sugar subsidy program transferred an average of $2 million to each domestic sugar grower from consumers and taxpayers.[53] These programs do not exist to keep small-time farmers in business. Author Brian Finegan writes that

$235 million worth of subsidies each year go to enrich just 158 sugar farmers, with one family alone receiving $60 million in subsidies from the program annually.[54] As sugar growers take money from taxpayers and consumers, they are destroying one of the country's environmental landmarks, the Florida Everglades.[55]

Green Scissors also blames peanut, cotton, and tobacco programs for encouraging farmers "to use large amounts of pesticides and over-farm their land."[56] Fertilizer use would decline with the removal of farm subsidies. Painter and Young found that eliminating farm subsidies would reduce nitrogen leaching from fertilizer use—the major cause of water pollution for many areas—by 46 percent in North Carolina's coastal plain.[57] Jonathan Tolman studied six farm states and estimated that the elimination of subsidies would result in a 35 percent decline in chemical use per acre and a 29 percent reduction for fertilizers.[58] The fact that there are numerous environmental benefits from removing subsidies is not theoretical. New Zealand farmer and farm political leader Brian Chamberlin documented a few of these benefits when his country eliminated subsidies during the 1980s:

> The result of removing subsidies is often cell intensive farming . . . which is beneficial to the environment. My proof does not lie in theory, but in practice. New Zealand, the only country in the world in recent years to have built up a system of high subsidies and then to have removed them, has found that to be the case. Since the removal of subsidies we have seen a much better balance between supply and demand, less farming of marginal and fragile land, a reduction in the overall use of chemicals and fertilizer, the return of much marginal land to forest, natural or commercial, and much more emphasis on cost control and quality rather than quantity in our produce.[59]

Even when agricultural subsidies are tailored with the specific goal of helping the environment, unintended consequences can cause them to go awry. Ethanol is a substitute for gasoline, with 95 percent of its supply coming from corn.[60] It is subsidized through tax credits and a reduction in the federal fuel excise taxes, because politicians and ethanol producers maintain it improves air quality by reducing the pollutants from gasoline. The subsidy amounts to nearly 54 cents per gallon of ethanol.[61] The majority of that money goes to agribusiness behemoth Archer Daniels Midland (ADM). ADM produces 60 percent of the country's ethanol and receives over $400 million each year in subsidies.[62] Still, shifting that money to a major corporation might be worthwhile if the ethanol subsidies actually helped clean the air, but it is not clear that they do.

The United States General Accounting Office found that removing ethanol subsidies would "slightly increase carbon monoxide emissions . . . but slightly reduce emissions of ozone precursors."[63] It concluded, "Available evidence . . . indicates that the ethanol tax incentives have had little effect on the environment,"[64] and by removing the ethanol program "little change in air quality or global environmental quality would be expected."[65] Henderson cites a study by the former chief of California's Air Resources Board. The chief explained that while carbon monoxide would be reduced with ethanol substituting for gasoline, other emissions would rise. Gasohols such as ethanol reduce carbon monoxide by 25 percent more than gasoline, but produce 50 percent more hydrocarbons and 15 percent more nitrogen oxide.[66] As a form of outdoor air pollution, nitrogen oxide is more likely to cause harm to humans than carbon monoxide. In addition, it is the interaction of hydrocarbons and nitrogen oxide that creates ground-level ozone, which is also more harmful to humans than carbon monoxide.[67] Hence, the ethanol subsidies may be making air quality worse for humans. As was noted in chapter three, the increased ethanol production could also overwhelm the spent grain market, leaving breweries with nowhere to send their spent grains but to the local landfill. Environmental benefits from ethanol are dubious at best, but the excessive cost to the environment in the form of more corn production, which leads to damaged habitat, increased fertilizer runoff, and higher pesticide use, are clear.

The Conservation Reserve Program (CRP) is another subsidy that transfers funds to farmers with the intention of helping the environment. CRP was introduced under the 1985 Food Security Act to pay farmers for retiring lands from production that were likely to suffer erosion. As additional goals, it encouraged farmers to replant native grasses, improve water quality, and provide wildlife habitat.[68] In return for not producing crops in certain areas, farmers in CRP get paid on a per acre basis. In 1996, the program encompassed 36 million acres and cost about $1.8 billion for an average payment of $50 per acre. This is nearly twice what the farmers would have received from farming or renting the land.[69] By 2001, CRP received only $1.65 billion in funding, but the program will be bumped up under President Bush's budget to $1.78 billion in 2002.[70]

It turns out, perhaps not surprisingly, that paying farmers not to grow a crop also produces unintended consequences that harm the environment.

As lands are removed from production, farmers make up the yield in two ways. First of all, as happened in similar set-aside programs in decades past, they convert other land previously not under till into agricultural lands. Philip Gersmehl, a geographer from the University of Minnesota, conducted a five-year study. He found "for every eroding acre a farmer idles, another farmer—or sometimes the same one—simply plows up nearly as much additional erosion-prone land."[71] On the Great Plains, Gersmehl noted that farmers had been paid to set aside 17 million acres for CRP, yet the total land cultivated in the region only dropped by 2 million acres.[72] Even when land is taken out of production for CRP, the environment can suffer. Output of agriculture remains unchanged from CRP because farmers not only make up the difference by converting more land but also by increasing agricultural intensity elsewhere. Gardner writes that farmers cover the difference by putting marginal land into the reserves and then substituting additional labor and chemicals on their other property to make up the difference.[73] This means more pesticides and fertilizer.

Subsidies might hurt the environment, but it is argued they help small farmers stay in business. Not true. The best place to look for the effect of removing subsidies on agriculture is New Zealand. In 1984, the government announced it would eliminate subsidies completely in 1987, and then largely did so. Subsidies comprised only 1 percent of New Zealand farm income in 1998.[74] During the five years from 1981 to 1986, the number of people employed in agriculture declined by 9,000. With the elimination of subsidies in 1987, the agricultural workforce declined over the next seven years to 1994 by another 10,500. The number of people actually employed on farms, however, went up from 1981 to 1994—increasing by about 7,000 workers.[75] Much of the reduced agricultural workforce came from corporations that created fertilizer or other products encouraged by subsidies. Those corporate jobs were transferred to other sectors as the resources previously devoted to farm subsidies returned to the general economy and allowed new investment in business and jobs elsewhere. New Zealand's economy has grown stronger since ending its massive subsidy system.

Most federal subsidies go to farmers who do not need them. The top 10 percent of U.S. farmers receive 61 percent of the billions of subsidy dollars.[76] According to Finegan, farmers with annual sales in excess of

$100,000 receive 65 percent of all agricultural subsidies, while those with annual sales less than $10,000 receive only 5 percent of the subsidies.[77] Even the subsidies that go to the, "small farmers," are largely unwarranted as many of the small-income farms are actually hobby farms run by well-to-do folks in search of tax breaks. There is no reason for either these hobby farmers or the larger, rich farms to receive subsidies. A 2001 GAO report broke down the subsidies and found:

> In recent years, over 80 percent of farm payments have been made to large- and medium-size farms, while small farms have received less than 20 percent of the payments. For example, in 1999 (the latest year for which data were available), large farms—the 7 percent of farms nationwide with gross agricultural sales of $250,000 or more—received about 45 percent of the payments. The 17 percent of farms that are medium-sized (gross sales between $50,000 and $249,000) received 41 percent of the payments. The remaining 14 percent of the payments was shared by the 76 percent of farms that are small (gross sales under $50,000). Small farms substantially outnumber medium and large farms, but because payments are generally based on volume of production, the average payment of small farms that received payments was much less. In 1999, these small farms, on average, received payments of about $4,141. In contrast, large farms received payments averaging about $64, 737, while medium-sized farms received average payments of about $21, 943.[78]

Ironically, the subsidies may well be leading to the demise of small farms in the U.S. Elizabeth Becker of the *New York Times* reports the subsidies are often used as capital for large operators who want to expand their holdings by buying out smaller neighbors.[79] The GAO uncovered that because subsidies often lead to higher prices for buying or leasing farmland, young people wishing to start their own small farms are discouraged from agriculture.[80]

Brian Finegan summarizes the ludicrous institutions set up by farm subsidies:

> The U.S. has been subsidizing farmers almost since its inception. We pay farmers not to grow crops in areas with abundant rainfall, while building multi-billion dollar dams so that other farmers can grow water-loving crops in deserts. Then we pay them not to grow crops, too. The taxpayers freely loan farmers money at subsidized rates. If they do not repay the loans, we loan them more money. We pay them to overproduce food and

then pay foreigners to take it off our hands. In the name of helping family farmers, huge agribusinesses—corporate food processors and exporters—receive billions of dollars in subsidies.[81]

The United States subsidizes farming one way or another via numerous programs to the tune of $47 billion each year.[82] Since 1996, $71 billion in direct payments were made to the farming sector.[83] In 2000, direct payments reached an all-time annual high of $20 billion.[84] It is the taxpayers and the environment that are paying for these welfare checks. The Organization for Economic Cooperation and Development estimates that from 1986 to 1994, subsidies cost U.S. taxpayers $370 billion.[85] This was practically enough money to buy all of the farmland in the country over that period.[86] The federal government is running a program that traps farmers in a form of welfare and it is harming more than agriculture's reputation. As true in any industry, successful farmers and agriculturalists would probably be better off without subsidies. Some experts estimate world market prices for crops would rise 25 to 50 percent if export subsidies and import barriers were removed.[87]

It has been said that there are two parties in the United States: the stupid party and the evil party. With the 2002 Farm Bill, the parties showed that they can get together under the banner of bipartisanship to do something both stupid and evil. At an estimated cost of $190 billion over ten years, President Bush signed a bill passed by the Democratically-led Senate to increase the payments to mostly large farm operations by $83 billion more than the existing programs.[88] Indeed, with communism's fall in Eastern Europe, U.S. agriculture is one of the few places where five-year plans still exist. And the 2002 plan was a doozy.

The examples presented here indicate that farmers and the environment would prosper with the removal of agricultural subsidies. The perverse incentives created by the Endangered Species Act and the Federal Estate Tax demonstrate that unintended consequences can arise from good intentions. If these programs were doomed to institutional failure, what sorts of institutions are destined for success? We will attempt to address this question in the concluding chapter that follows.

8.

INSTITUTIONS FOR
A BOUNTIFUL HARVEST

It is too easy to be a critic. It is much more difficult to offer constructive suggestions on how to improve the institutional structure governing agriculture and environmental stewardship. We do not pretend to have all of the answers for an agricultural policy that will continue to feed the world, provide for society's varying environmental demands, and deal with unseen future challenges. We do believe, however, that lessons can be learned from the mistakes covered in chapter seven as well as the successes heralded in the other chapters to construct some basic rules for building institutions that aid all players in flourishing within a dynamic world.

The various taxes, regulations, and subsidies discussed in chapter seven are part of the institutional structure in which farmers, environmentalists, and other landowners work. The institutions discussed in that chapter demonstrated how unintended consequences can arise from poorly designed institutions which then work against environmental goals regardless of stated intent.[1] The programs in chapters seven are by no means a comprehensive list of the failed and successful institutions that affect the everyday decisions of agrarians, consumers, environmentalists, and policy makers. The changing marketplace, the rules established by government agencies and courts, and the cultural milieu all play important parts in how agrarians act and react to the demands of consumers and other interests. This includes a variety of complex, and often conflicting, viewpoints. What then can be done to produce institutions that allow agriculture to meet varying goals? What institutions will encourage agrarians to meet the consumer demand for environmental stewardship?

153

Controlling the marketplace or changing the cultural milieu are both tasks and goals beyond the capability and expertise of these authors. Hence, we limit the recommendations that follow to the institutional structure set up by government bodies through public policy. But in making these policy recommendations, we cannot ignore that changing markets and culture affect the governing institutions. When certain goods grow scarce and others grow more bountiful, the market evolves. As this book has demonstrated, wealthier people with more disposable income have led to an increased demand for environmental goods. In the future, new demands will create new markets and old demands might fade away. Culture changes, too. A once dark and frightening forest of monsters for mankind to conquer and cut down became a pristine woodland of nature to preserve as wilderness areas grew scarce and knowledge changed.

These changes in cultures and markets will affect and change viewpoints on what constitutes the best solution available in any given place and time. This means solutions will vary from place to place and from one era to the next. Successful institutions need to adapt to these changes. For that to happen, the institutions must focus on setting up dynamic rules of the game so participants can roll with the punches. With that in mind, we draw four lessons from the preceding chapters for fomenting institutions that will help eco-entrepreneurs in agriculture to blossom.

1. Eliminate government subsidies.
2. Protect existing property rights and recognize new forms of property rights.
3. Don't let the sticks crowd out the carrots.
4. Innovate institutions, don't institutionalize innovations.

Eliminate governmental subsidies.

Uncle Sam is the typical bachelor uncle. He means well, but his good intentions often get him into trouble. Government subsidies start under the auspices of a helping hand. But subsidy-makers usually overemphasize one goal without adequately considering how it impacts other worthy goals. Failing to consider the whole picture, institutions are developed that create benefits for an action, but ignore the relative costs of the action.

Subsidies never advocate general goods such as quality air. They focus on bolstering a specific path to achieving air quality, such as ethanol use. But what happens when the solution advocated turns out to be unsuccessful or, even worse, counterproductive? By advocating one solution at the expense of others, subsidies have reduced competition. They have homogenized the situation so that everyone has been hit with the same bad solution.

Subsidy standards are set in a vacuum. They do not consider the side effects that might arise from the subsidy. Interconnectedness between goods in an economy is ignored. After all, what is clean air? What is the best mix of hydrocarbons, carbon monoxide, oxygen, and other gases and particulate matter?

Subsidies also fail in keeping up with the changing landscape. Significant congressional review of a subsidy program is done every four or five years at most. Even with that review, few subsidies are ever eliminated because they create entrenched interests by encouraging people to invest their time and capital in the subsidized good. Those who benefit from the subsidy will fight for its continuation long after it has any value. This is why certain subsidies have persisted for almost 200 years and have a good chance of persisting for 200 more.

While the subsidies don't change or go away, new technologies and knowledge do emerge every day that make the subsidies obsolete or counterproductive. These advances alter what we understand to be the best course of action when securing environmental quality and agricultural well being. Without constant review and subsequent repeal, subsidies push for the solutions of the past, not those of the present and future. This hinders agriculture's evolutionary process.

Subsidies discourage investment in finding new technologies. They discourage innovation by leading investors to chase government dollars instead of the dollars attached to discovering a new market niche or tool. The inability of subsidies to change at the speed of the marketplace leaves them vulnerable to the havoc of unintended consequences. Additional funding for ideas that appear to offer a working solution today makes it harder for new, innovative ideas to take hold. Thus, subsidies reduce competition in the marketplace of ideas by favoring one solution over others. Eco-entrepreneurs succeed by identifying new opportunities created by changing times and technologies. At the same time, subsidies fight them every step of the way to continue employing the vestigial techniques of antiquity. If Mother Nature had relied on subsidies, human beings would likely still have sloping foreheads and tails.

Subsidies redistribute research funds from the solutions of tomorrow to catalyze the solutions of yesterday. There is no such thing as a free lunch.

The money to subsidize one sector means some other sector has funds taken from it. Funding for government subsidies comes through taxes that give rise to possible environmental harms like those from the estate tax. Governments must raise funds to provide for things like national defense, the court system, and local police, but raising taxes to subsidize programs in the name of helping the environment and agriculture when those programs accomplish the opposite of what they intend is a double slap in the face.

One of the greatest challenges for ecological agrarians is getting their innovative ideas to gain widespread adoption. It could be argued subsidies accomplish this by helping eco-entrepreneurs to grow their businesses. The problem is that in practice, subsidies do not end up in the hands of eco-entrepreneurs. As pointed out in chapter seven, subsidies go to established interests. By advocating businesses and systems already in place, the subsidies make it more difficult for eco-entrepreneurs to get their innovations into the marketplace, not less so. If an idea is indeed a success like those ideas mentioned in this book, it will spread on its merits as increased profits from increased consumer demand push it forward. Predator Friendly Wool is growing by doing a good job, not by accepting government subsidies that promote its product over traditional wool.

Subsidies do not necessarily preclude eco-entrepreneurs from succeeding, but they do make the job more difficult. The ranching for wildlife agrarians in chapter five have to compete with the subsidy of cheap, almost free, hunting on public lands. The subsidized hunting makes it impossible for ranching for wildlife operators to compete on price. They cannot enter the low-end market to offer reasonable hunts to low income hunters, because the public lands are practically giving away such hunts. Ranching for wildlife operations can only compete with the public lands for hunters who care about the quality of a hunt. In essence, there are no fast food operations in ranching for wildlife because the government is giving the hamburgers away. With the mass market closed to them, ranching for wildlife operations must settle for high-end clients who want to buy a steak. This may seem like a blessing, but in fact, it limits the opportunities for expansion of ranching for wildlife's environmental success. After all, there are a lot more people who buy fast food than dine at steak houses.

Protect existing property rights and recognize new forms of property rights.

Good environmental institutions rely on the tried-and-true phrase, "think globally, act locally." As the history of agriculture in chapter one indicates, the world and its available technologies are constantly changing, creating a near infinite amount of dispersed knowledge. Situations in different places change from one moment to the next. Trying to synthesize this information of time and place to make decisions about how resources should be used and how their use affects the environment, agriculture, and the world at large is a daunting task at best. Nobel-laureate economist Friedrich Hayek realized this when he wrote:

> If we can agree that the economic problem of society is mainly one of rapid adaptation to changes in the particular circumstances of time and place, it would seem to follow that the ultimate decisions must be left to the people who are familiar with these circumstances, who know directly of the relevant changes and of the resources immediately available to meet them.[2]

Keeping pace with the constant changes in technologies and demands requires that agriculturalists have the power to make decisions about the resources that they govern. Property rights give eco-entrepreneurs that power.

The power provided by property rights guides owners to act as proper stewards for two reasons. First, property holders have pride in their possessions because it is a personal reflection on themselves. Their property is generally the product of their own hard work or that of their ancestors. The reason a person owns land, a car, or a telephone is because that person earned the money to pay for it or because someone passed down wealth earned from previous labors.

Second, property rights encourage us to take care of our belongings because they lose value if we do not. A rancher who overgrazes his property lowers its value. It has less worth for him because there isn't enough grass to feed his cattle, the aesthetics are poor, and the environmental quality is diminished. It is less valuable to potential buyers for the same reasons.

A property right places decision making power over a resource in the domain of the property owner, allowing him or her to capitalize on time and place specific knowledge to consume the resource, invest in expanding it, or simply preserve it for a rainy day. For instance, the owner of a tree farm can cut down trees for sale, plant seedlings for future use, or let already planted trees remain growing. Decisions made by property owners tend to be superior, because property owners have the best information about the resources in question and they are the ones most directly affected by the costs and benefits of any particular decision regarding the property.

To encourage proper stewardship and optimal decisions, property rights must encompass three components.[3] First of all, a property right must be definable. Without clear boundaries of what belongs to the property owner, there is no way to enforce one's decision making power. Others must know what belongs to whom. Until property rights in wildlife were established through ranching for wildlife, the wildlife was unowned and landowners had no incentive to improve habitat for the animals. The decision makers for the habitat and the decision makers for the animals were not linked.

Second, a property right must be defensible. If one's neighbors can condemn property without giving the owner anything in return, the owner has no incentive to conserve or invest in the future.[4] He or she has no guarantee of reaping the fruits of labor when it is harvest time. This makes immediate plundering of the property before the value is taken by others the best strategy. Investment and conservation are not options. If any hunter can walk onto a property and shoot the wildlife, there is no incentive for the landowner to conserve the animals. Similarly, if the World Wildlife Fund (WWF) cannot defend against the use of its logo by potato farmers who do not meet its standards, it has no incentive to invest in building a brand name.

Finally, a property right needs to be transferable. If a property holder cannot sell his property to others, he or she cannot reap the benefits of trading their labor for other goods. Consumers who value the landholder's property for different reasons (including its environmental benefits) will not be able to express their values, because they will not be able to reward the property owner by purchasing the fruits of the property or the property itself. As Ronald Coase illustrated, transferability ensures that resources go to their highest valued use.[5] Without exchange, changing values cannot be expressed. If the breweries in chapter three could not sell their spent grain, the recycling operation with dairy farmers could never get off the ground.

Innovative property rights have helped the entrepreneurs in this book to succeed. Defining property in game animals and waterways produced incentives for the environmental improvements discussed in chapter five. Defensible property protected the Milesnicks and the Teller Preserve's ability to manage for environmental goods. It defended against others dictating how they should manage their land. In chapter three, transferable property moved resources from low valued use to higher valued use.

Despite these successes, incomplete property rights and the condemnation of property still keep some entrepreneurs from reaching their full potential. Under current institutions, the Oregon Water Trust can lease water rights, but if it chooses to purchase them, the rights must be turned over to the state. This prevents the trust from using the water rights as capital for exchanges that might allow it to trade one water right for an even better water right. As landholders are forced to divide up their properties to pay off the heavy burden of the estate tax, the large properties required for effective ranching for wildlife programs are threatened with segmentation.[6] The Endangered Species Act obstructs entrepreneurs from investing in endangered species to grow healthy populations that might have a commercial value, because no one is allowed to own the species. Worse yet, encouraging such species on an individual's property gives the government legal authority to restrict land use practices and thereby reduces property value.

For property rights to work, they must be definable, defensible, and transferable. Once people have property rights to a resource, they can act on their localized knowledge. They can make a decision based on the time and place specific information provided in a given instant.[7] And they can weigh the costs and benefits of that decision.

Don't let the sticks crowd out the carrots.

Discussing the ESA in 1995, Michael Bean of Environmental Defense said, "The Endangered Species Act has generated controversy because it has been all stick and no carrot."[8] The primary means for accomplishing the goal of the act has been to penalize bad actions instead of rewarding good ones. Carrots rely on positive incentives, while sticks rely on negative incentives.

There are significant problems with using punishment for bad behavior instead of rewarding good behavior. The first is information. Those causing harms or engaging in improper behavior benefit by concealing information from the punishment enforcer. People do not wish to suffer punishment. They are unlikely to reveal harms they are committing. At the same time, people committing harms are the ones most likely to know that harm is occurring. Under the Endangered Species Act, it is tantamount to the "shut up" part of "shoot, shovel, and shut up." Landowners with endangered species are wise to keep the information to themselves, so that the U.S. Fish and Wildlife Service does not regulate their land. Lueck emphasizes that the victim of any penalties under the ESA, the landowner, has a distinct information advantage over the enforcer of the law, the U.S. Fish and Wildlife Service.[9] The stick encourages the landowner to conceal that information. This makes the job of species protection increasingly difficult.

On the other hand, rewarding people for doing good encourages positive environmental action to emerge. The marketplace focuses on rewarding good stewards with financial benefits. For instance, Becky Weed helps protect endangered and non-endangered predators, because consumers reward her with a premium on her wool. All of the green market opportunities in chapter two rely on capturing increased revenues, the carrot of the marketplace, by providing information to consumers.

But the Endangered Species Act works as a big stick that prevents markets from forming in endangered species and shutting down information exchange. With secure property rights, endangered species could be protected by landholders who hope to profit off of their environmental value to society. With the ability to charge to view rare species or even hunt them on a sustainable basis, property owners are given an incentive to protect endangered species habitat and encourage the growth of endangered populations just as property rights and an open market encouraged wildlife habitat in the ranching for wildlife programs.[10] Landowners would have an incentive to share information about endangered species on their property, because they could use it as a selling tool for their goods. Conservation groups could pay landholders for protecting habitat and endangered species or they could buy the property from the landowners to manage themselves. At present, the looming stick of the Endangered Species Act crowds out these opportunities, because no landowner wants to admit that they have rare species on their property and so conservation groups often don't know where to look to make such transactions.

One might ask where the difference lies between the carrot and the government subsidies above that we advise avoiding as a solution. After all, subsidies dangle a carrot to get individuals to act in a certain manner. But the key difference is that government subsidies are carrots that ignore the importance of property rights and the varying importance of different environments.

Government subsidies are paid for with taxes taken from the citizenry at large and then spent through a proliferation of statutes that do not take quality into account the way that a private organization does. When the Nature Conservancy offers to buy a piece of property for habitat or offers to pay a farmer to place a conservation easement on their property, it first determines the importance of the habitat. Even with an annual budget of $500 million, the Nature Conservancy is limited in the amount of funds it can spend and it wants to get the most bang for its environmental buck.[11]

In comparison, the vast funding of a federal subsidy program does not force the government to make frugal choices. The government is not as accountable for its actions as a private organization. If the Nature Conservancy spends its money poorly, its members can withdraw funding. Citizens upset with government spending can attempt to elect an official who will be thriftier, but they do not have the option of withholding their tax dollars. These perverse attributes, among others,[12] mean that federal subsidies will often be less effective or even counterproductive carrots as the examples in the last chapter demonstrated. While the Nature Conservancy researches each property to preserve only the most environmentally significant lands, a federal subsidy like the Conservation Reserve Program accepts habitat of varying levels of importance, including lands better left to farming. It could be argued that all habitat preservation is good, but that ignores the fact that we must employ some land to provide our food. It makes sense that properties better adapted to farming than providing environmental amenities remain in farming. Federal subsidies do not distinguish between the two, because the costs of buying lands are dispersed. The Nature Conservancy, in contrast, does distinguish between habitats, because it more directly feels the impact of preservation costs through diminished funds. This is why the private carrot and the public carrot are not one in the same.

Still, for both public and private entities, the carrot inspires an atmosphere of cooperation that the stick does not. The stick creates an incentive to avoid punishment. The carrot creates an incentive to work towards a mutual goal. Everyone wins with the carrot. Those giving the reward see their goal forwarded and those receiving the reward are compensated for their actions. Under the stick, someone loses. As Deborah Kane, executive director of The Food Alliance,[13] said, "There is a realization . . . that the stick approach has had limited returns, and it has left a bad taste in peoples' mouths."[14]

Innovate institutions, don't institutionalize innovations.

The evolutionary successes shared in this book were the result of agricultural eco-entrepreneurs seeing a need and stepping forward to fill it. They were chasing the dollars available from the marketplace—the dollars from those willing to pay for environmental benefits. Still, the success stories on these pages should not be considered models to turn into government programs. Just because General Motors demonstrates how to build good cars does not mean the government should imitate them and get into the automotive business. Such thinking suffers from the same problems as subsidies. It ignores the fact that today's solution might not be the best answer tomorrow. The world and its environment are dynamic.[15]

The inability of a program institutionalized by the government to meet changing demands leads to programs that are eventually outdated, harmful, and protected for decades by entrenched interests. This does not mean that there is not an important role for the government to play. The government has a crucial role. It must be the arbiter. It must set up the rules of the game by which the entrepreneurs play. It must enforce the property rights and negotiated contracts to ensure the game is played by fair rules. At the same time, as the paragraphs below warn, the government must not become a player in the game.

Water markets offer an excellent example of the importance of government in creating institutions that set the rules of the game while at the same time exposing the folly of government acting as a player in the game. As chapter five demonstrated, the rules needed changing to allow organizations to buy water and keep it instream for environmental purposes. Without a rule change (an innovation of institutions), the environmental value of leaving water in situ for the recreation, aesthetics, and peace of mind that aquatic life and riparian ecosystems provide to people could not be recognized. Eco-entrepreneurs like the Oregon Water Trust could not buy water from farmers and ranchers. Agrarian entrepreneurs like Rocky Webb could not sell water to environmental groups. It was here that the government stepped in to change the rules of the game to allow for innovation.

The state of Oregon worked as an arbiter to establish the rules of the game. It dictated what could be traded, how trades could occur, when trades could happen, who could buy or sell water, and countless other rules that were prescriptive enough so that all parties involved knew what they could and could not do, but were flexible enough to allow entrepreneurs to take positive action. Organizations like the Oregon Water Trust jumped at the chance to play by these new rules. They went out and they began making innovative deals to help both agrarian interests and environmental interests. Other groups followed suit in Nevada, Washington, and Montana. At this point, the government should have stepped back, recognized it had done well, and rested until a dispute between players needed settling. Unfortunately, the government took a different route. Not satisfied with innovating institutions, it decided to start institutionalizing the innovations made by the private players. With the rules set, it entered the game and began buying water for environmental purposes in the same way that the Oregon Water Trust was doing.

In this case, the problem with the government entering the game as a market participant is two-fold. First, government agents do not have the same fiscal frugality that private participants like the water trusts do. If they make poor transactions, they do not go out of business. Instead, they ask for more money from the treasury the next year. Governments have deep pockets and are not as sensitive to price. Playing fast with their money, they drive up the price of water far above what the private entities were paying, making it more difficult for water trusts to survive and do a good job. If the water trust is run out of business, the entrepreneurial innovator who came up with the idea in the first place, and may well have come up with even better ideas, has been eliminated from the game. It is similar to the situation that keeps the ranching for wildlife entrepreneurs out of the low-end hunting market. The government runs them out of town with deep pockets and irreverence for what consumers demand.

The second problem with government agents entering the game is that as the market arbitrator, they can change the rules of the game. This places the government on unequal footing with the rest of the participants and creates a significant conflict of interest. It places the other participants on shaky ground and encourages them to take their investments where the playing field doesn't include the 500-pound gorilla that can change the rules for everyone at will.

When the government steps in to act as a participant in the marketplace, it all too often institutionalizes innovations. When the government acts as the rule-maker, it is innovating institutions. The importance of the rule-making role should not be downplayed, because no other entity has the power to enforce the rules of the game like the government. With the coercive power of the state behind it, only governments can make people play by the rules. That same coercive power is why it is so important the government isn't a player itself, because it cannot guarantee fairness when it has a stake in the game.

Even in making the rules of the game, the government needs to be careful to make sure that it is setting up institutions to help the market players to interact and contract with one another. It needs to create rules that reduce barriers, rather than increase them. In the discussion over spent grain and ethanol in chapter three, the government chose to set rules that were too strict and that caused interactive barriers between market participants. With an inflexible regime that hurt the environment and agriculture, the government, required an oxygenate additive in gasoline, which set a rule that prevented Conoco and Tosco from improving air quality through other avenues.

Eco-labeling, fee-hunting, water leases, and the fertilizing sewage technology of Sheaffer International are wonderful services. Such successes do not mean the government should start a nationwide campaign to set up fee-hunting programs or mandate Sheaffer technology in every community. In the future, photo-safaris and manure as alternative fuel may be the way to go. No one can know since tomorrow's solutions haven't been tested or even thought up yet. Hence, we should avoid institutionalizing today's solutions with the backing of government regulation and dollars. Instead, let the market, communities, and ecological agrarians chasing after the fruits of their labors be our guide into tomorrow. Let government set the rules by which they will play.

It is tempting to take a success and try to implement it everywhere, but the best way to copy success is to let entrepreneurs bring the idea to new areas. If an innovation offers a successful opportunity elsewhere, entrepreneurs will help that good idea flourish in the marketplace. If, on the other hand, the idea can't be replicated successfully, then there is a good chance that it shouldn't be. Good ideas don't need Uncle Sam's help to propagate, because market investment will follow a good idea and help it expand to greener pastures.

Conclusion

We believe institutions based on the suggestions offered here will provide for better stewardship by the agricultural sector while continuing to put food on our tables. They encourage a proliferation of solutions from the rich and diverse assembly that makes up humanity. They then let the consumer citizenry choose what works best while at the same time leaving the door open for different and more effective solutions in the future. Combining flexibility with rewards for good ideas and restraint on putting every egg into a single basket, these institutions rely on the ingenuity that lies in each and every one of us to deal with a changing world.

With that said, we admit once more that we do not have all of the answers to making agricultural businesses more profitable. We do not possess a magic wand to wave over agrarians to improve environmental performance. What we offer are rules of the game. These rules will enable eco-entrepreneurs who come up with the solutions to different problems to get their idea into the marketplace where the consumer citizenry can judge it a success or failure. Relying on an evolution of ideas, we trust the meritorious programs will survive while those lacking merit will fall to the wayside.

By no means will the entrepreneurial efforts in this book make the farmers or ranchers practicing them rich. Nor will they replace "traditional" farming and ranching as a way of life. But, as we have seen, agriculture is evolving. Eco-entrepreneurship could be the difference between survival and going belly up for many farmers and ranchers just like that lone milk cow was the difference between life and death so long ago. It is important, then to get the proper incentives in place via proper institutions. By ignoring or distorting the institutions of property, the marketplace, prices, and contracting, significant harm could be done to both agriculture and the ecosystems that depend on agricultural lands.

In essence, the game has not changed since the beginning of civilization. We live in an ever-more dynamic world. Agriculture is changing to solve problems, and with each solution new problems will no doubt arise. The experiences recounted in this book may provide some direction for agriculture's next step and institutions to guide the steps that follow. Ultimately, however, the future lies in the hands of agricultural entrepreneurs owning and caring for their land. Everything else will follow from these ecological agrarians.

Notes

Notes to Chapter One

1. Ehrlich 1968: i.
2. Ehrlich 1970: 23–25.
3. Diamond 1999: 150–151.
4. The barley grown and harvested in the United States today is not the little barley of the Native Americans, but barley brought over from Europe that originated in the wilds of the Fertile Crescent.
5. In addition to Diamond 1999, see Crosby 1986.
6. Vasey (1992: 23–25) offers a good account of differing theories on when and how agriculture arose.
7. Diamond 1999: 181. Murray (1970) offers a detailed account of the rise of agriculture through domesticated animals and plants in Europe from the beginnings in the Fertile Crescent to its spread into France around 2000 BC.
8. Diamond 1999: 261.
9. Ibid.
10. Heiser 1981: 2.
11. Diamond 1999: 234.
12. Heiser 1981: 19.
13. Malthus 1993.
14. Grigg 1982: 22.
15. Heiser 1981: 12.
16. Gras 1925: 191–192.
17. White 1970: 146.
18. White 1970: 19, 149.
19. Goklany 1998: 942. This is a significant drop in the change of productivity compared to the 200-fold figure Gras provides for Mesopotamia. We conjecture the difference lies in inferior measurement and record-keeping of the Mesopotamian figures compared to modern figures, characteristics unique to the region compared to average soil, or exaggeration on Gras's part.
20. Carrier 1923: 267–272.
21. Gras 1925: 192.
22. Carrier 1923: 267.
23. White 1970: 125.
24. Quoted in White 1970: 132.
25. Cochrane 1993: 127.
26. Goklany 2001.
27. McClelland 1997.
28. Carrier 1923: 263.
29. Vasey 1992: 245.
30. Cochrane 1993: 126.
31. Grigg 1982: 178.

32. White 1970: 174–187.

33. Goklany 2001.

34. The Old Testament's proclamation that land must rest every seventh year may be a reference to the fallow system, though Gras (1925: 26) has doubts.

35. Oerke et al. 1994

36. See Heiser (1981) for an entire volume on how better seeds led to our modern civilization.

37. Braudel 1979: 132.

38. This number is calculated using numbers on personal consumption expenditures from the Bureau of Economic Analysis cited in the *World Almanac and Book of Facts 2000*. For reference, see Famighetti (1999: 130). Much of the expense of food consumption today includes luxury food eating. In the modern world, food is as much entertainment as it is nutrition when money is spent on expensive restaurants and exotic foods. Thus, the expense for basic nutrition is well overestimated by these numbers and our modern expenditures on food as a percent of income could be even less.

39. Lomborg 2001: 62.

40. Braudel 1979: 133.

41. Braudel (1979: 135) lists the hours of labor required to purchase a quintal or hundredweight of wheat at 100 hours of work. On page 134, he writes that the average family of four consumes 12 quintals of grain per year.

42. Braudel 1979: 132.

43. This is calculated by dividing the 1,200 hours worked each year to provide for grain needs by the 3,650 pounds of bread eaten per year.

44. Cox and Alm 1999: 43.

45. Lomborg 2001: 62.

46. Cochrane 1993: 127.

47. For discussion of institutions, rising wealth, and higher survival rates effects on population, see Norton (2002) and Becker and Barro (1988).

48. Singer 1999: 22; Avery 2000: 52.

49. UN Population Division 2001: v; Lomborg 2001: 47. In the last decade, population has mimicked the UN's lower estimates more often than its mid- or high-level estimates.

50. Lomborg 2001: 109.

51. Sen 1999: 6.

52. Avery 2000: 148–167.

53. Carrier 1923: 10.

54. Cochrane 1997: 282.

55. U.S. Environmental Protection Agency 2000b.

56. Coursey 1992.

57. Speech to International Consumers for Civil Society on November 29, 1999 in Seattle, Washington.

58. Goklany 2001:12.

59. Borlaug 2000.

60. Cochrane 1993: 128.

61. Avery 2000: 32.

62. Gardner 2001: 97.

63. Miller 2001: 106.

64. Cochrane 1993: 127.

65. Beattie 2001: 16.

66. Miller 2001: 108.

67. Beattie 2001: 17.

68. Staley 2001: 67.

69. U.S. Department of Agriculture 2000c: 6.

70. Gardner 2001: 88.

71. Gardner 2001: 88.

72. Heimlich et al. 2000: 13–14. An acre-treatment is the number of acres treated times the number of pesticide treatments applied.

73. Kelly 1997.

74. Avery 2000: 195; Brady and Weil 1999: 156, n12.

75. See Brady and Weil (1999: 691–696) for discussion of conservation tillage and its advantages.

76. Avery 2000: 45.

77. Bowditch 1999: 50.

78. DeGregori 2001: 143.

79. Quoted in Hill 2000.

80. Long 2000: 58.

81. DeGregori 2001: 143.

82. Famighetti 2000: 136.

83. Cochrane 1993: 160.

84. Famighetti 2000: 136.

85. Cochrane 1993: 160.

86. Gras 1925: 301.

87. Staley 2000: 71. The shift in practical terms is probably even larger, as many of the small farms today are hobby farms and tax write-offs. Hence, the average size of real, working farms is probably even bigger.

Notes to Chapter Two

1. For a discussion of the Endangered Species Act, see Chapter Seven.

2. The group was originally called the Oregon Rivers Council.

3. The Council helped pass the Oregon Omnibus National Wild and Scenic River Act, which listed 40 rivers and 1,500 river miles with a wild and scenic rivers designation.

4. Telephone conversation with Dan Kent, managing director, Salmon-Safe, Eugene, Oregon, 13 September 2001.

5. Hawley 1999.

6. Email from Dan Kent, managing director, Salmon-Safe, Eugene, Oregon, 15 October 2001.

7. Pacific Rivers Council 2001.

8. Telephone conversation with Dan Kent, managing director, Salmon-Safe, Eugene, Oregon, 13 September 2001.

9. Salmon-Safe 2001.

10. Hawley 1999.

11. Blend and van Ravenswaay 1999.

12. Woodbury 2000.

13. Ibid.

14. Weed 2000. Many other sheepherders use guard dogs to protect their sheep from bears and mountain lions. Becky Weed's operation may go that route someday, too.

15. Chasteen 1999. In 2001, conventional wool averaged between 20¢ and 75¢ per pound, with a typical price near 30¢ per pound.

16. These prices were downloaded from Becky Weed's website at http://www.lambandwool.com on October 13, 2001. Sweaters range from $89 to $188, blankets from $88 to $185, and hats from $19 to $33.

17. Email from Becky Weed, Thirteen Mile Ranch, Belgrade, Montana, 30 November 2001.

18. Email from Becky Weed, Thirteen Mile Ranch, Belgrade, Montana, 21 November 2001.

19. Quoted in Wilkinson 1997: 3.

20. Quoted in Davis 2000.

21. Quoted in Wilkinson 1997: 3.

22. Email from Becky Weed, Thirteen Mile Ranch, Belgrade, Montana, 21 November 2001.

23. Ibid.

24. Quoted in Nature Conservancy 2000.

25. Nature Conservancy 2000.

26. Ibid.

27. Schmidt 2001.

28. Ibid.

29. This information is taken from a copy of the agreement between Defenders of Wildlife and rancher Jim Winder. Mailing from Jenny Neeley, Defenders of Wildlife Southwest Office, Tucson, Arizona, 27 June 2001.

30. Animal Damage Control Services, now called Wildlife Services, subsidizes the destruction of predators with USDA funding. For a discussion of Wildlife Services, see Grewell 2002a.

31. Quoted in Bendrick 1999.

32. Quoted in Chasteen 1999: 21.

33. Quoted in Wilkinson 2000a: 22.

34. Quoted in Chasteen 1999: 23.

35. The guidelines used by Salmon-Safe are available for download from its website at http://www.salmonsafe.org.

36. Email from Dan Kent, managing director, Salmon-Safe, Eugene, Oregon, 19 September 2001.

37. Salmon-Safe 2001.

38. Telephone conversation with Jeff Dlott, Board of Directors member for Protected Harvest, Watsonville, California, 6 December 2001.

39. Fulmer 2001.

40. Quoted in Kades 2001.

41. WWF et al. 2001; Fulmer 2001.

42. Quoted in WWF et al. 2001.

43. McCallum and Brown 2001.

44. Telephone conversation with Jeff Dlott, Board of Directors member for Protected Harvest, Watsonville, California, 6 December 2001.

45. Quoted in Green 2001.

46. Telephone conversation with Jeff Dlott, Board of Directors member for Protected Harvest, Watsonville, California, 6 December 2001.

47. Kades 2001.

48. Quoted in McCallum and Brown 2001.

49. Telephone conversation with Lori Sandman, executive director, Environmental Quality Initiative, Strausstown, Pennsylvania, 4 November 2001.

50. Ibid.

51. Sandman 1999.

52. Quoted in Dionis 2000.

53. Food Alliance 2001a; Kane 2001.

54. Food Alliance 2001b.

55. Food Alliance 2001c.

56. Moscatello 2001.

57. Food Alliance 2001c.

58. Food Alliance 2001d.

59. Food Alliance 2001c.

60. Quoted in Chasteen 1999: 20.

61. Chasteen 1999.

62. At the beginning of Chapter Seven, we discuss some of these new regulations as well as how regulations can often have the opposite effect of their intended goal.

63. Quoted in Chasteen 1999: 19.

64. Email from Becky Weed, Thirteen Mile Ranch, Belgrade, Montana, 4 December 2001.

65. Quoted in Walker 2000: 3.

66. Chasteen 1999.

67. Ibid.

68. Quoted in Nixon 1998.

69. Email from Becky Weed, Thirteen Mile Ranch, Belgrade, Montana, 4 December 2001.

70. Quoted in Chasteen 1999: 20.

71. For notions of risk-avoidance in less developed parts of the world, see Scott (1976).

72. For more information on Free Farmed, see http://www.freefarmed.org.

73. Quoted in Humane Society of the United States (HSUS) 2001.

74. Telephone conversation with Craig Miller, Defenders of Wildlife Southwest Office, Tucson, Arizona, 19 June 2001.

75. Chasteen 1999: 21.

76. See Chapter Seven for further expansion of government's static nature in dealing with problems.

77. Chasteen 1999: 21.

78. Kane 2001.

79. DeGregori 2001: 160.

80. Fulmer 2001.

81. Gardner 2001: 94.

82. Chasteen 1999: 16.

83. Anderson and Leal 1997: 77.

Notes to Chapter Three

1. Beer Institute 2001.

2. Ibid.

3. Wornson 1989.

4. U.S. Environmental Protection Agency (EPA) 2001b.

5. Quoted in Truini 2001.

6. Quoted in O'Malley 1997.

7. Chandler 1986.

8. Telephone interview with Dan Dwyer, byproducts manager for Miller Brewing Company, 12 July 2001.

9. Perrin and Klopfenstein 2000.

10. Ibid.

11. Telephone interview with Steve Rockhold, Special Products Manager for Coors, 16 August 2001.

12. Desrochers 2000.

13. Miller 1998.

14. Telephone interview with Charlie Staff, executive director of the Distillers Grains Technology Council, 17 August 2001.

15. Beer Institute 2001.

16. Quoted in Truini 2001.

17. Ibid.

18. Ibid.

19. Ibid.

20. Telephone interview with Peter Rochefort, environment specialist with Molson, 6 June 2001.

21. Ibid.

22. Telephone interview with Barry Robinson, farm nutritionist with Great Northern Livestock, 16 July 2001.

23. Telephone interview with Peter Rochefort, environment specialist with Molson, 6 June 2001.

24. Quoted in Strachan 1995.

25. McCarthy and Tiemann 2001.

26. Ibid.

27. Ibid.

28. Chevron 1997.

29. California Air Resources Board 2000.

30. Powers et al. 2001.

31. Telephone interview with Charlie Staff, executive director of the Distillers Grains Technology Council, 17 August 2001.

32. Ibid.

33. From internal company documents received via email on 15 August 2001 from Dan Dwyer of Miller Brewing.

34. Quoted in Grewell 2001: 1.

35. For a history of the Chesapeake Bay Agreements between the states surrounding the Bay, the city of Washington, D.C. and the Environmental Protection Agency, see www.chesapeakebay.net.

36. Gilmore et al. 2000. The agreement is available for download from http://www.chesapeakebay.net/agreement.htm.

37. Gilmore et al. 2000: 5.

38. Fabre 2000.

39. This is quoted from a company brochure entitled "An Idea Whose Time Has Come."

40. Grewell 2001.

41. Puckett 2001.

42. Telephone interview with Dave Mullan, engineer for Sheaffer International, Naperville, Illinois, 12 April 2001.

43. Progressive Farmer 2001.

44. Email from Dave Mullan, engineer for Sheaffer International, Naperville, Illinois, 15 February 2001.

45. Grewell 2001: 13.

46. Goodman 1999a, 1999b, 1999c.

47. EPA 2000b.

48. From internal company documents received via facsimile on 13 November 2000 from Sheaffer International, Limited.

49. From fact sheet received in Sheaffer International press packet received via mail on 10 October 2000.

50. Grewell 2001.

51. From internal company documents received via facsimile on 30 November 2000 from Sheaffer International, Limited.

52. Quoted in Trice 2000.

53. Quoted in Fabre 2000.

54. This is quoted from the Sheaffer International Brochure "An Idea Whose Time Has Come." No date of publication or footnote is available.

55. Telephone interview with Dave Mullan, engineer for Sheaffer International, Naperville, Illinois, 12 April 2001.

Notes to Chapter Four

1. U.S. Department of Agriculture (USDA) 2000b.

2. Environmental Defense 2001a.

3. Ibid.

4. The authors would like to acknowledge and thank Kris Kumlien for his valuable research assistance in helping tell the Milesnicks' story.

5. Quoted in Montana Stockgrower 2001.

6. Montana Stockgrower 2001.

7. Wilkinson 2000b.

8. Quoted in Montana Stockgrower 2001.

9. Conversation with Tom Milesnick, Belgrade, Montana, 12 September 2001.

10. Ibid.

11. Ibid.

12. Conversation with Mary Kay Milesnick, Belgrade, Montana, 12 September 2001.

13. Quoted in Wilkinson 2000b.

14. Quoted in Montana Stockgrower 2001.

15. Ibid.

16. The story of the Milesnicks' access problem is a Tragedy of the Commons as articulated by Hardin (1968) and solved by property rights. The evolution from open access to property systems in general is expounded on in Rose (1991).

17. Quoted in Renewing the Countryside 2001.

18. Rebuffoni 1992.

19. Ibid.

20. Dayton Hudson Corporation is one of the largest U.S. general merchandising retailers and is ranked among the top companies on the Fortune 500 list. The company's holdings include retail outlets such as Target, Mervyns, and Marshall Fields.

21. Quoted in King 2001.

22. Quoted in Renewing the Countryside 2001.

23. Quoted in Rebuffoni 1997.

24. Quoted in PERC 1998.

25. Quoted in American Farmland Trust 2000.

26. Quoted in http://www.tellerwildlife.org/concept.html. Downloaded on July 27, 2000.

27. Telephone conversation with Diane Boyd, Teller Wildlife Refuge Executive Director, Teller Wildlife Refuge, Montana, 18 December 2001.

28. Quoted in http://www.tellerwildlife.org/. Downloaded on July 27, 2000.

29. Telephone conversation with Amy Moneteith, Education Program Director, Teller Wildlife Refuge, Montana, 14 December 2001.

30. Telephone conversation with Diane Boyd, Teller Wildlife Refuge Executive Director, Teller Wildlife Refuge, Montana, 18 December 2001.

31. Ibid.

32. Ibid.

33. Telephone conversation with Amy Moneteith, Education Program Director, Teller Wildlife Refuge, Montana, 14 December 2001.

34. Ibid.

35. Ibid.

36. Telephone conversation with Diane Boyd, Teller Wildlife Refuge Executive Director, Teller Wildlife Refuge, Montana, 18 December 2001.

37. Ibid.

38. Ibid.

39. Quoted in Nature Conservancy 2001.

40. Ibid.

41. Quoted in Kinsella 2001.

42. Ibid.

43. Ibid.

44. Quoted in http://nature.org/wherewework/northamerica/states/massachusetts/misc/leclair.html/. Downloaded on January 28, 2002.

45. Quoted in Kinsella 2001.

46. Telephone conversation with John Curelli, Executive Director, FARM Institute, Martha's Vineyard, Massachusetts, 24 August 2002.

Notes to Chapter Five

1. Quoted in Grewell and Peck 1999a: 13.

2. Ibid.

3. Leal and Grewell 1999: 30.

4. Beattie 2001: 8; Gardner 2001: 83.

5. Kelly 1997: 9.

6. U.S. General Accounting Office (GAO) 1994: 5–6.

7. Quoted in Environmental Defense 1999.

8. See subsidies section in chapter seven for a discussion of the Conservation Reserve Program.

9. Examples such as the Federal Estate Tax, the 1973 Endangered Species Act, and agricultural subsidies are discussed in chapter seven.

10. Leopold 1991: 202.

11. Quoted in Grewell and Peck 1999a: 16.

12. Quoted in Leal and Grewell 1999: 20.

13. Ibid.

14. The eight state programs (California, Colorado, Nevada, New Mexico, Oklahoma, Oregon, Utah, and Washington) are described in more detail in Leal and Grewell (1999).

15. No more than 14 of the tags may be used for bucks. Information provided via email communication from Sarah Edmonds, California Fish and Game Department, 8 April 1999.

16. Telephone conversation with Bill Burrows, Burrows Ranch, Tehama County, California, 9 February 1999.

17. Ibid.

18. Telephone conversation with Jeff Weinstein, Corning Land and Cattle Company, Tehama County, California, 8 February 1999.

19. Ibid.

20. Leal and Grewell 1999: 23.

21. Email communication from Sarah Edmonds, California Fish and Game Department, 8 April 1999.

22. Fitzhugh 1989.

23. Gooch 1998.

24. Leal and Grewell 1999: 23.

25. U.S. Department of the Interior et al. 1997: 24. The Burrows hunt does not include equipment, but it does include food, drink, hunting access, guides, and tags.

26. The Boone and Crockett Club is a hunting organization founded by Theodore Roosevelt in 1887. Like the Safari Club, it keeps hunting records of trophy-size animals. For antlered animals, scoring is based on the animal's rack.

27. Quoted in Leal and Grewell 1999: 14.

28. Leal and Grewell 1999: 16.

29. Telephone conversation with Bill Burrows, Burrows Ranch, Tehama County, California, 9 February 1999.

30. Colorado's formal program, Raching for Wildlife, provides ranching for wildlife with the generic name by which all programs are known.

31. Telephone conversation with Dave Menagetti, Twin Peaks Ranch, Las Animas County, Colorado, 9 August 1999.

32. Leal and Grewell 1999: 24.

33. Telephone conversation with Dave Menagetti, Twin Peaks Ranch, Las Animas County, Colorado, 9 August 1999.

34. Quoted in Stalling 1999: 70.

35. Telephone conversation with Dave Menagetti, Twin Peaks Ranch, Las Animas County, Colorado, 9 August 1999.

36. Telephone conversation with David Buschena, economics professor, Montana State University, 17 July 2001 regarding data from Colorado Division of Wildlife, Big Game Hunting Statistics, Denver, CO (1993–1997).

37. Hess and Wolf 1999.

38. Freddy et al. 1991: 340.

39. Telephone conversation with Ty Ryland, manager, Forbes Trinchera Ranch, Costilla County, Colorado, 12 July 2001.

40. Telephone conversation with Ty Ryland, manager, Forbes Trinchera Ranch, Costilla County, Colorado, 12 July 2001.

41. Email communication from Jerry Apker, Colorado Division of Wildlife, 18 July 2001.

42. Hess and Wolf 1999: 36.

43. Telephone conversation with Ty Ryland, manager, Forbes Trinchera Ranch, Costilla County, Colorado, 12 July 2001.

44. Grewell and Peck 1999b: 15.

45. Ibid.

46. Fears 1996: 2.

47. Fears 1996.

48. Telephone conversation with David Stevens, landowner, Grant County, Washington, 19 August 1999.

49. Leal and Grewell 1999: 38.

50. Leal and Grewell 1999: 17.

51. Glick et al. 1998.

52. Quoted in Stalling 1999: 73.

53. Telephone conversation with Dale Spencer, landowner, Alton and Paunsaugunt units, Utah, 12 August 1999.

54. Hess 2001a.

55. Ibid.

56. See comment on back cover of Leal and Grewell (1999).

57. Another major point of contention over water is availability for growing urban areas. Whole books have been written on the problem of providing water for both agriculture and municipalities. For possible solutions to water scarcity, see Anderson and Snyder (1997).

58. Quoted in Oregon Water Trust (OWT) 1996.

59. Quoted in OWT 1996.

60. Laatz 1994.

61. Solley, Pierce, and Perlman 1998.

62. Quoted in Kiewel 2001: 1.

63. Solley, Pierce, and Perlman 1998.

64. U.S. Department of Agriculture (USDA) 1997.

65. See Landry (1998) for different water laws and their use to protect instream flows.

66. Telephone conversation with Bob Hanson, Montana Farm Bureau Federation, Whitehall, Montana, 6 October 1998.

67. Telephone conversation with Ginny Larson, Missoula, MT, 8 October 1998.

68. Ibid.

69. Telephone conversation with Bob Hanson, Montana Farm Bureau Federation, Whitehall, Montana, 6 October 1998.

70. Telephone conversation with Ted Eady, Sisters, Oregon, 1 October 1998.

71. Ibid.

72. Telephone conversation with Bob Hanson, Montana Farm Bureau Federation, Whitehall, Montana, 6 October 1998.

Notes to Chapter Six

1. The term "biotechnology" has been used for nearly 100 years to describe any process of applying scientific knowledge to the development of biological organisms. In that regard, biotechnology could be used to describe many traditional types of plant breeding. More recently, the term has come to represent, among lay people, only the most advanced techniques of genetic engineering and recombinant DNA. In this chapter, I use biotechnology in this latter, more specific meaning.

2. Prakash 2000.

3. James 2001.

4. U.S. Department of Agriculture (USDA) 2001.

5. National Research Council (NRC) 1987.

6. NRC 1989: 13.

7. See, for example, NRC 1989; Royal Society et al. 2000; World Health Organization (WHO) 1991; American Medical Association (AMA) 2000.

8. Rissler and Mellon 1996.

9. Oerke et al. 1994.

10. U.S. Environmental Protection Agency (EPA) 2001a.

11. AGBIOS 2001.

12. Carpenter and Gianessi 2001; Dunn 1998.

13. Carpenter and Gianessi 2001; USDA 2000a.

14. Ferber 2000.

15. Progressive Farmer 2000.

16. Gianessi and Silvers 2001.

17. Carpenter and Gianessi 2001.

18. Kilman 2001.

19. Ferber 1999.

20. Carpenter and Gianessi 2001.

21. Smith and Leonard 2001.

22. Perks 2001; Pray, Ma, Huang, and Qiao 2001.

23. Ismael, Yousouf, Bennett, and Morse 2001. "Farm level impact of Bt cotton in South Africa." *Biotechnology and Development Monitor*, No. 48 (December).

24. Datta 2000.

25. Tu et al. 2000.

26. See the two studies in Losey, Rayor, and Carter 1999 and Hansen and Obrycki 2000.

27. Miflin 2000: 19.

28. Pimentel and Raven 2000; Taylor 1999; Pleasants, Hellmich, and Lewis 1999; Hellmich, Lewis, and Pleasants 2000a; Hellmich, Lewis, and Pleasants 2000b.

29. EPA 2000a.

30. EPA 2000a; Sears et al. 2001; Stanley-Horn et al. 2001; Hellmich et al. 2001; Oberhauser et al. 2001; Pleasants et al. 2001; and Zangerl et al. 2001.

31. Monarch Watch 2001.

32. Pimentel and Raven 2000.

33. Miflin 2000.

34. Stotzky 2000; Saxena and Stotzky 2001; Saxena and Stotzky 2000.

35. Doyle 1999; Pray, Ma, Huang, and Qiao 2001.

36. USDA 2001.

37. Carpenter and Gianessi 2001.

38. Benbrook 2001; Hin, Schenkelaars, and Pak 2001; Parrott 2001.

39. See, for example, Fernandez-Cornejo and McBride 2000; Carpenter and Gianessi 2001; Benbrook 2001; Hin, Schenkelaars, and Pak 2001.

40. Hin, Schenkelaars, and Pak 2001. See Hin, Schenkelaars, and Pak 2001 and Environmental Defense 2001b concerning the environmental effects of glyphosate.

41. Ag-West Biotech 2001. On cotton, see Fernandez-Cornejo and McBride 2000; Carpenter and Gianessi 2001; Coble 1999.

42. Avery 1995.

43. Crosson 1995; McGraw and Comis 2000.

44. Koskinen and McWhorter 1986; see Lehmann and Pengue 2000 and American Soil Association (ASA) 2001 for the connection between conservation tillage and Roundup Ready soybeans in Argentina and the United States, respectively.

45. Cook 1999; Carpenter and Gianessi 1999.

46. Squire et al. 1999.

47. Trewavas and Leaver 2001.

48. Crawley et al. 2001

49. Tilman 1999.

50. DeGregori 2001; quotation from p. 84.

51. Guerinot 2001.

52. McHughen 2000.

53. Lopez-Bucio et al. 2000; de la Fuente et al. 1997.

54. Lopez-Bucio et al. 2000; de la Fuente et al. 1997.

55. Conway and Toenniessen 1999.

56. Takahashi 2001.

57. Wong 2001; Golovan et al. 2001.

58. Fetrow 1999; Butler 1999.

59. Main, Roka, and Noss 1999; Vitousek et al. 1997.

60. McNeely and Scherr 2001.

61. Statistics from Goklany 1998; Goklany 1999.

62. Mann 1999; Conway and Toeniessen 1999.

63. UN Development Program 2001.

64. Goklany 1999.

65. McNeely and Scherr 2001.

66. Quoted in Mann 1999: 313.

67. Moffat 2000.

68. Goklany 1999.

69. See Baker, Zambryski, Staskawicz, and Dinesh-Kumar 1997; Moffat 2001; Beachy 1999; Gianessi and Silvers 2001.

70. Gianessi and Silvers 2001; Westwood 2001.

71. Cockcroft 2001.

72. Miflin 2000; Pennisi 2001; Moffat 2000.

73. Benbrook 2001; Hin, Schenkelaars, and Pak 2001.

74. Sayler 2000.

75. Vörösmarty, Green, Salisbury, and Lammers 2000.

76. Pennisi 2001; Miflin 2000.

77. Alvim et al. 2001.

78. Alia, Sakamoto, and Murata 1998; Jaglo-Ottosen et al. 1998.

79. Sommerville and Briscoe 2001.

80. Kasuga et al. 1999.

81. Frommer, Ludewig, and Rentsch 1999.

82. Raven, Evert, and Eichhorn 1992; Smirnoff and Bryant 1999.

83. Apse et al. 1999; Zhang and Blumwald 2001; Frommer, Ludewig, and Rentsch 1999.

84. National Public Radio 2001.

85. Times of India 2001; Hargrove 2001.

86. Organic Trade Association (OTA) 2001; Soil Association 2000.

87. Kain 1996.

88. On copper sulfate, see National Library of Medicine 2001; Gianessi 1993. On rotenone, see Kain 1996. On pyrethrum, see EPA 1999; Kain 1996; Trewavas 2001.

89. Gianessi 1993.

90. Trewavas 2001.

91. Trewavas 2001.

92. MacKerron et al. 1999.

93. MacKerron et al. 1999.

94. Trewavas 2001.

95. Organization for Economic Cooperation and Development (OECD) 2001; OECD 2001; Bizily, Rugh, and Meagher 2000; Ryoo et al. 2000; de Lorenzo 2001.

Notes to Chapter Seven

1. Goodman 1999a: A1.

2. EPA 2000b.

3. Morriss, Yandle, and Meiners 2001.

4. Lomborg 2001: 198.

5. Bowles 2001; Jehl 2001; Welch 2001

6. Kennedy 1999: 12.

7. Jenner 1998.

8. In relation to the historical context in which this book began, institutions have played pivotal roles in the technological advances that have occurred. This is especially true of the rapid advancement made in the last 200 years. For a discussion of the importance of institutions in this advancement, see Brenner (1976). He argues that England's institutions led it to earlier success in agricultural, capitalist, and industrial advancement when compared to the institutions of France; though arguably he mistakenly identifies the French system as offering more complete freedom and property rights for peasants. The fact that French peasants could not transfer their property and were heavily taxed created severe restrictions on property and personal freedom. The right of transfer is nearly as important to the value of property and

capitalist advancement as the ability to defend property against others taking from it. In addition, high taxes imposed by lord and state lowered personal freedom for peasants by tying peasants' hands in what decisions could be made. With these two aspects of French institutions in consideration, English institutions of property and freedom were actually the more complete.

9. Saxton and Thornberry 1998: 2.

10. Saxton and Thornberry 1998: iii.

11. Quoted in Saxton and Thornberry 1998: 32.

12. Quoted in Environmental Defense 1995.

13. U.S. General Accounting Office (GAO) 1994.

14. Adler 1999.

15. U.S. House 2000: H4158.

16. Greene et al. 2000.

17. Ibid.

18. By a vote of 54 to 44, the U.S. Senate voted down a permanent repeal of the estate tax on June 12, 2002 (Hulse 2002).

19. Endangered Species Act of 1973, Pub. L. No. 93-205, 87 Stat. 884 (1973).

20. Four Republicans voted against the bill: Robin L. Beard, Jr. (Tenn.), H.R. Gross (Iowa), Earl F. Landgrebe (Ind.), and Robert Price (Texas).

21. Congressional Quarterly 1973.

22. Sugg 1993a: 22.

23. Lueck 2000: 70.

24. Sugg (1993: 35–36) lists several sources that contend the legislative intent was not to include habitat modification on private land as a *take*.

25. Lueck 2000.

26. Huthcherson 1976.

27. *Tennessee Valley Authority v. Hill*, 437 U.S. 153, *160.

28. *Babbitt v. Sweet Home Chapter of Communities for a Great Oregon*, 515 U.S. 687. In this case, the Secretary of the Interior's definition of harm was challenged. The Secretary had devised a regulation in 1981 that defined a harm as "an act which actually kills or injures wildlife. Such act may include significant habitat modification or degradation where it actually kills or injures wildlife by significantly impairing essential behavioral patterns, including breeding, feeding, or sheltering." This definition is laid out in the federal code 50 C.F.R. § 17.3.

29. Quoted in Sugg 1997: 13.

30. Ibid.

31. Wilcove et al. 1996: 3.

32. Stroup 1995: 5.

33. Stroup 1995: 5.

34. Sugg 1993b.

35. Stroup 1995: 5.

36. Quoted in Sugg 1993b: A12.

37. Anderson and Leal 1997: 109-111.

38. Quoted in Anderson and Leal 1997: 110.

39. Ibid.

40. Lueck 2000: 109.

41. Four hundred and thirty-seven colonies is the highest concentration of woodpecker colonies in the data. Lueck and Michaels' data comes from the U.S. Forest

Service and a North Carolina State University survey of over 400 North Carolina landowners.

42. U.S. Fish and Wildlife Service 2001.

43. Gordon et al. 1997. The United States even lists species to the ESA not found on U.S. soil.

44. The U.S. Fish and Wildlife Service has tried to fix these poor incentives with the "safe harbors" program initiated in 1995. This program sets a baseline of species on the landowners' property and then ensures that no additional land use restrictions will be put on the property if additional endangered species choose to settle there. It falls under section 10 of the Act relating to habitat conservation plans. There are three problems with "safe harbors." First, there is a legal question of whether the program meets the statutory requirements of the ESA. If challenged in court, the program could be struck down, and landowners who entered would find themselves strapped with restrictions for the extra species on their property. Second, there is little guarantee that the Fish and Wildlife Service won't change its mind in the future and begin to regulate where the safe harbors program has identified endangered species to exist. Due to the program's questionable legality, it seems unlikely the landowners involved would be compensated for such a shift in the administering agency's philosophy. Third, the program still imposes costs on landowners by requiring that they engage in certain activities to encourage endangered species on their property. Safe harbors allows landowners to avoid more binding restrictions in the future by engaging in some voluntary measures up front. In light of the first two problems with safe harbors, the costs imposed by the third become more substantial since there is no guarantee the voluntary efforts will provide sufficient reward in the future.

45. For further discussion on the environmental harm of subsidies, including nonagricultural ones, see Leal and Meiners (2002).

46. Gardner 2001: 83.

47. Roberts 1999: 55.

48. Landry 2002.

49. Green Scissors 2001: 26.

50. Bovard 2001: 18.

51. GAO 2000: 3.

52. GAO 2000: 5.

53. Finegan 2000: 25.

54. Finegan 2000: 132.

55. Landry 2002.

56. Green Scissors 2001: 4.

57. Painter and Young 1994: 456.

58. Cited in Gardner 2001: 87.

59. Chamberlin 1996: 69.

60. GAO 1997: 5.

61. GAO 1997: 10.

62. Finegan 2000: 124.

63. GAO 1997: 6.

64. GAO 1997: 5.

65. GAO 1997: 14.

66. Cited in Henderson 1992: 58.

67. U.S. Environmental Protection Agency (EPA) 1996: V49–V53; Lomborg 2001.

68. Hess 2001b.
69. Carey 1996: 6.
70. Kelley 2001.
71. Cited in Carey 1996: 6.
72. A report from the GAO found that the Acreage Reduction Programs that pre-ceded CRP failed similarly. From 1982 to 1985, participating wheat farmers idled 55 million acres under the programs, while nonparticipating wheat farmers increased cultivation on 35 million acres (Bovard 1989: 100).
73. Gardner 2001: 85.
74. Brooke 1999: 1.
75. Chamberlin 1996: 85.
76. Becker 2001.
77. Finegan 2000: 119.
78. GAO 2001: 2.
79. Becker 2001.
80. GAO 2001: 6, 25–27.
81. Finegan 2000: 117.
82. Brooke 1999.
83. Green Scissors 2001: 4.
84. GAO 2001: 1.
85. Bovard 1995.
86. Finegan 2000: 118.
87. Avery 1998.
88. Allen 2002.

Notes to Chapter Eight

1. For more examples of institutions that harm the environment, see Leal and Meiners (2002).
2. Hayek 1945:524.
3. The importance of each of these components and of property as an institution is outlined in Anderson and Leal (2001) and Epstein (1995).
4. The most common scenario of neighbors banding together to take all or a portion of another neighbor's property is in the form of a government action. For in-stance, government condemnation of property is regular practice for building roads. The conscription of certain property rights is also common in endangered species en-forcement as demonstrated in chapter seven. While landholders are generally com-pensated for land taken to build a road, they are not compensated for the property restrictions put in place by the Endangered Species Act.
5. Coase 1960.
6. As we noted in chapter seven, the estate tax is in the process of being phased out over the course of the next decade, before it is reinstated in 2011.
7. Norton (1998: 51) found that "environmental quality and economic growth rates are greater in regimes where property rights are well defined than in regimes where property rights are poorly defined."
8. Quoted in Environmental Defense 1995.
9. Lueck 2000.
10. Cultivation of endangered species to provide income for increased habitat is currently being used in Africa with much success. Trophy hunting and photography

expeditions have turned the large animals into an asset and led to increasing herd populations as the local citizens are granted property rights to manage the elephants as they see fit. For further discussion on this, see Anderson and Leal (2001: 153–155) and Grewell (2002a).

11. For further discussion on the perverse incentives created by a lack of accountability for federal purse strings and those incentives' effect on the environment, see Grewell (2002b).

12. Another perversion created by the system is the federal government's need to be perceived as fair and equal to all parties. While fairness and equity of government treatment are important virtues that should be encouraged, they create problems when it comes to selectivity of purchase. The government must set specific standards that it enforces in all cases. It cannot discriminate. But this makes it difficult for the government to discriminate between less and more important environmental projects when it starts to allocate subsidy dollars. Thus, we end up with programs where specific standards determine the eligibility for the dollars, not performance or quality of the good being purchased. Government subsidies generally set up shop and say that if you meet this minimum, come sign up, and we'll pay you. Private carrots go out and seek the projects or lands they actually think will make a difference and only make offers to those individuals rather than to every single American with ten acres to hoe.

13. See chapter two for discussion about The Food Alliance and eco-labeling.

14. Quoted in Lee 2000.

15. See Postrel (1998: 111–146) for rules that work in a dynamic world.

References

Adler, Jonathon H. 1999. The Anti-Environment Estate Tax: Why the "Death Tax" Is Deadly for Endangered Species. *CEI On Point Policy Brief.* April 20.

AGBIOS. 2001. Global Status of Approved Genetically Modified Plants. On-Line Database maintained by Agriculture & Biotechnology Strategies, Inc. Downloaded from http://64.26.172.90/agbios/dbase.php?action=Synopsis on November 29, 2001.

Ag-West Biotech. 2001. GM Crops Reduce Pesticide Use. *Agbiotech Bulletin* 9(6). August. 1–2.

Alia, H., H. A. Sakamoto, and N. Murata. 1998. Enhancement of the Tolerance of Arabidopsis to High Temperatures by Genetic Engineering of the Synthesis of Glycinebetaine. *The Plant Journal* 16(2). October. 155–161.

Allen, Mike. 2002. Bush Signs Bill Providing Big Farm Subsidy Increases. *Washington Post.* May 14. A1.

Alvim, Fátima C., Sônia M. B. Carolino, Júlio C. M. Cascardo, Cristiano C. Nunes, Carlos A. Martinez, Wagner C. Otoni, and Elizabeth P. B. Fontes. 2001. Enhanced Accumulation of BiP in Transgenic Plants Confers Tolerance to Water Stress. *Plant Physiology* 126(3). July. 1042–1054.

American Farmland Trust. 2000. The Campaign for America's Farmland. Downloaded from http://www.writingforagooD.C.ause.com/aft.pdf on July 27, 2000.

American Medical Association. 2000. Report 10 of the Council on Scientific Affairs (I-00): Genetically Modified Crops and Foods. Chicago, IL: American Medical Association. Available at http://www.ama-assn.org/ama/pub/article/2036-3604.html.

American Soybean Association. 2001. *Conservation Tillage Study.* Chicago: American Soybean Association, November.

Anderson, Terry L., and Donald R. Leal. 1997. *Enviro-Capitalists: Doing Good While Doing Well.* Lanham, MD: Rowman & Littlefield Publishers.

————. 2001. *Free Market Environmentalism.* Revised edition. New York: Palgrave.

Anderson, Terry L., and Pamela Snyder. 1997. *Water Markets: Priming the Invisible Pump.* Washington, D.C.: Cato Institute.

Apse, Maris P., Gilad S. Aharon, Wayne A. Snedden, and Eduardo Blumwald. 1999. Salt Tolerance Conferred by Overexpression of a Vacuolar Na_+/H_+ Antiport in *Arabidopsis. Science* 285(5431). August 20. 1256–1258.

Ausubel, Jesse H. 2000. The Great Reversal: Nature's Chance to Restore Land and Sea. *Technology in Society* 22 (August). 289–301.

Avery, Dennis T. 1995. *Saving the Planet with Pesticides.* Indianapolis, IN: Hudson Institute.

————. 1998. What the Future May Hold for US Farmers. *Bridge News.* Downloaded from http://www.hudson.org/averydoc1.htm on March 24, 1999.

185

————. 2000. *Saving the Planet with Pesticides and Plastic.* Second edition. Indianapolis, IN: Hudson Institute.

Baker, Barbara, Patricia Zambryski, Brian Staskawicz, and S. P. Dinesh-Kumar. 1997. Signaling in Plant-Microbe Interactions. *Science* 276(5313). May 2. 726–733.

Beachy, Roger N. 1999. Testimony before the U.S. Senate Committee on Agriculture, Nutrition and Forestry. October 6. Available at http://www.senate.gov/~agriculture/Hearings/Hearings_1999/bea99106.htm.

Beattie, Bruce R. 2001. The Disappearance of Agricultural Land. In *Agriculture and the Environment,* ed. Terry L. Anderson and Bruce Yandle. Stanford, CA: Hoover Press. 1–23.

Becker, Elizabeth. 2001. Far From Dead, Subsidies Fuel Big Farms. *New York Times.* May 14.

Becker, Gary, and Robert J. Barro. 1988. A Reformulation of the Economic Theory of Fertility. *Quarterly Journal of Economics* 103. February. 1–25.

Beer Institute. 2001. *Annual Beer Almanac.* Washington, D.C.: Beer Institute.

Benbrook, Charles. 2001. Troubled Times amid Commercial Success for Roundup Ready Soybeans. AgBioTech InfoNet Technical Paper Number 4. May 3. Available at http://www.biotech-info.net/troubledtimes.html.

Bendrick, Mary Lou. 1999. Wolf Country Beef: How One Rancher Coexists with Wolves. *Mountainfreak.* Fall. 32–33.

Bizily, Scott P., Clayton L. Rugh, and Richard B. Meagher. 2000. Phytodetoxification of Hazardous Organomercurials by Genetically Engineered Plants. *Nature Biotechnology* 18(2). February. 213–217.

Blend, Jeffrey R., and Eileen O. van Ravenswaay. 1999. Measuring Consumer Demand for Ecolabeled Apples. *American Journal of Agricultural Economics* 81(5). 1072–1078.

Borlaug, Norman. 2000. Ending World Hunger: The Promise of Biotechnology and the Threat of Anti-science Zealotry. *Plant Physiology* 124. October. 487–490.

Bovard, James. 1989. *The Farm Fiasco.* San Francisco, CA: ICS Press.

————. 1995. Kill Farm Subsidies Now. *Washington Post.* October 13.

————. 2001. Candy Mountain: It's Time to Take Sugar Producers Off Welfare Rolls. *Investor's Business Daily.* June 18. 18.

Bowditch, Sherwood. 1999. America's Farmers: No Good Deed Goes Unpunished. *ALEC Policy Forum.* December. 50–52.

Bowles, Scott. 2001. Oregon Farmers Deep in Water Fight: Drought Heats Protests against Protection of Endangered Fish. *USA Today.* July 9. 3A.

Brady, Nyle C., and Ray R. Weil. 1999. *The Nature and Properties of Soils.* Twelfth edition. Upper Saddle River, NJ: Prentice-Hall, Inc.

Braudel, Fernand. 1979. *The Structures of Everyday Life: The Limits of the Possible.* Vol. 1, *Civilization and Capitalism, 15th–18th Century.* New York: Harper and Row.

Brenner, Robert. 1976. Agrarian Class Structure and Economic Development in Pre-Industrial Europe. *Past and Present* 70. February. 30–74.

Brooke, James. 1999. Down and Out in Rural Canada. *New York Times.* December 18.

Butler, L. J. 1999. The Profitability of rBST on U.S. Dairy Farms. *AgBioForum* 2(2). 111–117.

California Air Resources Board. 2000. Health and Environmental Assessment of the Use of Ethanol as Fuel Oxygenate. Report to the California Environmental Policy Council in Response to Executive Order D-5-99. Downloaded from http://www-erd.llnl.gov/ethanol on September 14, 2001.

Carey, John. 1996. Old MacDonald Had a Boondoggle. *Business Week*. March 18.

Carpenter, Janet, and Leonard Gianessi. 1999. Herbicide Tolerant Soybeans: Why Growers Are Adopting Roundup Ready Varieties. *AgBioForum* 2(2). 65–72.

———. 2001. *Case Studies in Benefits and Risks of Agricultural Biotechnology: Roundup Ready Soybeans and Bt Field Corn*. Washington, D.C.: National Center for Food and Agricultural Policy. January.

Carrier, Lyman. 1923. *The Beginnings of Agriculture in America*. New York: McGraw-Hill Book Company, Inc.

Chamberlin, Brian. 1996. *Farming and Subsidies: Debunking the Myths*. Pukekohe, New Zealand: Euroa Farms Ltd.

Chandler, Paul. 1986. "Feeding Brewers Grain." Downloaded from http://www.inform.umd.edu/EdRes/Topic/AgrEnv/nd/Feeding_Brewers_Grain.htm on August 15, 2001.

Chasteen, Bonnie. 1999. Conscience, with a Price Tag: Eco-labels and Niche Brands Help Proven Stewards Stay on the Land. *Chronicle of Community* 3(2): 15–26.

Chevron. 1997. "Chevron Seeks Changes to Reformulated Gasolines." Press Release, Chevron Corporation Public Affairs Department. December 1.

Coase, Ronald H. 1960. The Problem of Social Cost. *The Journal of Law and Economics* 3(1). 1–44.

Coble, Harold D. 1999. Genetically Engineered Cotton. Paper presented at the conference Environmental Benefits and Sustainable Agriculture through Biotechnology. Georgetown University, Washington, D.C.. November 19.

Cochrane, Willard W. 1993. *The Development of American Agriculture: A Historical Analysis*. Second edition. Minneapolis: University of Minnesota Press.

Cockcroft, Claire. 2001. Yielding Results: Crop Engineering Is Improving Yields and Cutting Disease. *The Guardian*, London, UK. June 14. 21.

Congressional Quarterly. 1974. Endangered Species. In *1973 Congressional Quarterly Almanac*, ed. Mary Wilson Cohn. Washington, D.C.: Congressional Quarterly. 670–673.

Conway, Gordon, and Gary Toenniessen. 1999. Feeding the World in the Twenty-first Century. *Nature* 402(6761). December 2. C55–C58.

Cook, R. James. 1999. Testimony before the U.S. House of Representatives Subcommittee on Basic Research. October 5.

Coursey, Don. 1992. The Demand for Environmental Quality. John M. Olin School of Business, Washington University, St. Louis, MO. December.

Cox, W. Michael, and Richard Alm. 1999. *Myths of Rich & Poor*. New York: Basic Books.

Crawley, M. J., S. L. Brown, R. S. Hails, D. D. Kohn, and M. Rees. 2001. Transgenic Crops in Natural Habitats. *Nature* 409(6821). February 8. 682–683.

Crosby, Alfred W. 1986. *Ecological Imperialism: The Biological Expansion of Europe, 900–1900*. New York: Cambridge University Press.

Crosson, Pierre. 1995. Soil Erosion Estimates and Costs. *Science* 269(5223). July 28. 461–464.

Datta, Swappan K. 2000. A Promising Debut for Bt Hybrid Rice. Information Systems for Biotechnology *ISB News Report*. September. 1–3. Available at http://www.isb.vt.edu/news/2000/Dec00.pdf.

Davis, Seabring. 2000. "Hero for the Planet": Belgrade Woman Honored for Predator-Friendly Approach. *Bozeman Daily Chronicle*. February 21.

de la Fuente, Juan Manuel, Verenice Ramírez-Rodríguez, José Luis Cabrera-Ponce, and Luis Herrera-Estrella. 1997. Aluminum Tolerance in Transgenic Plants by Alteration of Citrate Synthesis. *Science* 276(5318). June 6. 1566–1568.

de Lorenzo, Victor. 2001. Cleaning Up behind Us: The Potential of Genetically Modified Bacteria to Break down Toxic Pollutants in the Environment. *EMBO Reports* 2(5). May. 357–359.

DeGregori, Thomas R. 2001. *Agriculture and Modern Technology: A Defense*. Ames: Iowa State University Press.

Desrochers, Pierre. 2000. "Cities and Industrial Symbiosis: How Unique Is Kalundborg." PERC Research Paper. PERC, Bozeman, MT.

Diamond, Jared. 1999. *Guns, Germs, and Steel*. New York: W.W. Norton and Company.

Dionis, Kim. 2000. "Green Milk" Successfully Test-Marketed at Mid-Atlantic Stores. *Penn State News*. May 31. Downloaded from http://www.psu.edu/ur/2000/greenmilk.html on October 22, 2001.

Dowd, P. F., R. L. Pingel, D. Ruhl, B. S. Shasha, R. W. Behel, D. R. Penland, M. R. McGuire, and E. J. Faron. 2000. Multiacreage Evaluation of Aerially Applied Adherent Malathion Granules for Selective insect Control and Indirect Reduction of Mycotoxigenic Fungi in Specialty Corn. *Journal of Economic Entomology* 93(5). October. 1424–1428.

Doyle, Ellin. 1999. Environmental Benefits and Sustainable Agriculture through Biotechnology. Executive Summary of the Ceres Forum. Georgetown University, Washington, D.C.. November 10–11, 1999.

Dunn, Sara M. 1998. From Flav'r Sav'r to Environmental Saver? Biotechnology and the Future of Agriculture, International Trade, and the Environment. *Colorado Journal of International Environmental Law & Policy* 9(1). Fall/Winter. 145–166.

Edwards, Stephen R. 1995. Conserving Biodiversity: Resources for Our Future. In *The True State of the Planet*, ed. Ronald Bailey. New York: The Free Press. 211–265.

Ehrlich, Paul. 1968. *The Population Bomb*. New York: Sierra Club Ballantine.

———. 1970. Looking backward from 2000 A.D. *The Progressive*. April. 23–25.

Environmental Defense. 1995. Unique Consensus on Endangered Species Incentives Embraces EDF Ideas. Environmental Defense News Release. July 31.

———. 1999. Environmental Defense Fund Will Pay Landowners to Protect Endangered Species. News Release. New York: Environmental Defense. December 7.

————. 2001a. Food for Thought: The Case for Reforming Farm Programs to Preserve the Environment and Help Family Farmers, Ranchers and Foresters. Report prepared by Environmental Defense, American Farmland Trust, Center for Science in Public Interest, Defenders of Wildlife, Environmental Working Group, and Trout Unlimited. Washington, D.C.: Environmental Defense. Report available at http://www.environmentaldefense.org/documents/160_foodforthought.pdf

————. 2001b. Glyphosate. *Environmental Defense Scorecard*, On-Line Database of Chemical Hazards. Downloaded from http://www.scorecard.org/chemical-profiles/hazard-indicators.tcl?edf_substance_id=1071%2d83%2d6 on November 29, 2001.

Epstein, Richard. 1995. *Simple Rules for a Complex World*. Cambridge, MA: Harvard University Press.

Fabre, Tracy. 2000. Shenandoah Valley Project: Manage Wastewater as a Resource Instead of Pollutant. *Water Online*. September 14.

Famighetti, Robert, ed. 1999. *The World Almanac and Book of Facts 2000*. Mahwah, NJ: Primedia Reference.

Fears, J. Wayne. 1996. Ranching for Wildlife: This Family Offers Quality Hunts and Improves Elk Habitat. *Progressive Farmer Online*. May. Downloaded from http://progressivefarmer.com/ruralsportsman/0596/ranching/index.html on June 3, 1999.

Ferber, Dan. 1999. GM Crops in the Cross Hairs. *Science* 286(5445). November 26. 1662–1666.

————. 2000. New Corn Plant Draws Fire from GM Food Opponents. *Science* 287(5457). February 25. 1390.

Fernandez-Cornejo, Jorge, and William D. McBride. 2000. *Genetically Engineered Crops for Pest Management in U.S. Agriculture: Farm-Level Effects*. Agricultural Economic Report No. 786. Washington, D.C.: USDA Economic Research Service, April.

Fetrow, John. 1999. Economics of Recombinant Bovine Somatotropin on U.S. Dairy Farms. *AgBioForum* 2(2). 103–110.

Finegan, Brian J. 2000. *The Federal Subsidy Beast*. Sun Valley, ID: Alary Press.

Fitzhugh, E. Lee. 1989. Innovation of the Private Lands Wildlife Management Program: A History of Fee Hunting in California. *Transactions of the Western Section of the Wildlife Society* 25. 49–59.

Food Alliance. 2001a. Frequently Asked Questions. Downloaded from http://www.thefoodalliance.org/faq.html on June 6, 2001.

————. 2001b. The Food Alliance and Environmental Defense Launch Sustainable Food Partnership. News Release. Downloaded from http://www.thefoodalliance.org/tfaedpr.html on June 6, 2001.

————. 2001c. Application Process and Fees. Downloaded from http://www.thefoodalliance.org/TFAApplication.pdf on June 6, 2001.

————. 2001d. The Food Alliance Guiding Principles. Downloaded from http://www.thefoodalliance.org/guiding.html on June 6, 2001.

Freddy, David J., Errol R. Ryland, and Richard M. Hopper. 1991. Colorado's Wildlife Ranching Program: The Forbes Trinchera Experience. In *Wildlife Production: Conservation and Sustainable Development*, ed. Lyle A. Renecker and R. J. Hudson. AFES Misc Pub 91-6. Fairbanks: Univ. of Alaska Fairbanks. 336–343.

Frommer, Wolf B., Uwe Ludewig, and Doris Rentsch. 1999. Enhanced: Taking Transgenic Plants with a Pinch of Salt. *Science* 285(5431). August 20. 1222–1223.

Fulmer, Melinda. 2001. Eco-Labels on Food Called Into Question. *Los Angeles Times*. August 26.

Gardner, B. Delworth. 2001. Some Issues Surrounding Land and Chemical Use in Agriculture. In *Agriculture and the Environment*, ed. Terry L. Anderson and Bruce Yandle. Stanford, CA: Hoover Press. 81–103.

Gianessi, Leonard P. 1993. Chemicals Seen Vital to Weed, Pest Control. *Forum for Applied Research and Public Policy* 8(3). Fall. 87–89.

Gianessi, Leonard P., and Cressida S. Silvers. 2001. *The Potential for Biotechnology to Improve Crop Pest Management in the US: 30 Crop Study*. Washington, D.C.: National Center for Food and Agriculture Policy, June.

Gilmore, James, Parris Glendening, Tom Ridge, Anthony Williams, Carol Browner, and William Bolling. 2000. *Chesapeake 2000: A Watershed Partnership*. June 28.

Glick, Dennis, David Cowan, Robert Bonnie, David Wilcove, Chris Williams, Dominick Dellasala, and Steve Primm. 1998. *Incentives for Conserving Open Lands in Greater Yellowstone*. Bozeman, MT: Greater Yellowstone Coalition.

Goklany, Indur. 1998. Saving Habitat and Conserving Biodiversity on a Crowded Planet. *Bioscience* 48(11). November. 941–953.

———. 1999. Meeting Global Food Needs: The Environmental Trade-Offs Between Increasing Land Conversion and Land Productivity. *Technology* 6(2–3). 107–130.

———. 2001. The Pros and Cons of Modern Farming. *PERC Reports* 19(1). March. 12–14.

Golovan, Serguei P., Roy G. Meidinger, Ayodele Ajakaiye, Michael Cottrill, Miles Z. Wiederkehr, David J. Barney, Claire Plante, John W. Pollard, Ming Z. Fan, M. Anthony Hayes, Jesper Laursen, J. Peter Hjorth, Roger R. Hacker, John P. Phillips, and Cecil W. Forsberg. 2001. Pigs Expressing Salivary Phytase Produce Low-Phosphorus Manure. *Nature Biotechnology* 19(8). August. 429–433.

Gooch, Bob. 1998. More on Hunts through "Ranching for Wildlife." *U.S. Hunting Report*. February.

Goodman, Peter S. 1999a. An Unsavory Byproduct: Runoff and Nutrient Pollution. *Washington Post*. August 1. Page A1.

———. 1999b. The Cost to the Bay: Permitting a Pattern of Pollution. *Washington Post*. August 2. Page A1.

———. 1999c. The Cost to the Bay: Who Pays for What Is Thrown Away? Impact of New Pollution Controls May Hinge on Liability for Manure. *Washington Post*. August 3. Page A1.

Gordon, Robert E., James K. Lacy, and James R. Streeter. 1997. Conservation under the Endangered Species Act. *Environment International* 23(3). 359–419.

Gras, Norman Scott Brien. 1925. *A History of Agriculture in Europe and America*. New York: Crofts.

Green Scissors. 2001. *Green Scissors 2001*, ed. Sarah Newport. Washington, D.C.: Friends of the Earth.

Green, Thomas A. 2001. Wisconsin Potatoes to Bear World Wildlife Fund Panda. Gempler's IPM Solutions. April. Downloaded from http://www.ipmalmanac. com/solutions/200104/panda.asp on October 13, 2001.

Greene, John, Tamara Cushing, Steve Bullard, and Ted Beauvais. 2000. Effect of the Federal Estate Tax on Non-Industrial Private Forest Holdings in the U.S. Extended abstract of paper for Fragmentation2000 Conference held September 17–20, 2000 at Annapolis, Maryland.

Grewell, J. Bishop. 2001. *Sheaffer International's BOOM Project*. Case Study SM-80. Stanford, CA: Stanford University Graduate School of Business. February.

———. 2002a. War on Wildlife. In *Government vs. the Environment*, ed. Donald R. Leal and Roger E. Meiners. Lanham, MD: Rowman & Littlefield.

———. 2002b. All Play and No Pay: The Adverse Effects of Welfare Recreation. In *Government vs. the Environment*, ed. Donald R. Leal and Roger E. Meiners. Lanham, MD: Rowman & Littlefield.

Grewell, J. Bishop, and Clint Peck. 1999a. Greenbacks for Bucks. *Montana Farmer Stockman*. December. 12–16.

———. 1999b. Making Wildlife Pay: Ranching for Wildlife Can Make Economic Sense for Landowners. Hunters Like It Too. *Colorado Farmer Stockman*. December. 15–16.

Grigg, David. 1982. *The Dynamics of Agricultural Change*. New York: St. Martin's Press.

Guerinot, Mary Lou. 2001. Improving Rice Yields—Ironing Out the Details. *Nature Biotechnology* 19(5). May. 417–418.

Hansen Jesse, Laura C., and John J. Obrycki. 2000. Field Deposition of Bt Transgenic Corn Pollen: Lethal Effects on the Monarch Butterfly. *Oceologia* 125. August. 241–248.

Hardin, Garrett. 1968. The Tragedy of the Commons. *Science* 162. December. 1243–1248.

Hargrove, Tom. 2001. China Announces Seawater Irrigation of GM Crops. *Planet Rice*. July 11. Available at http://www.planetrice.net/newsarchives/archivedetails. cfm?ID=783.

Hawley, Bob Jr. 1999. Farmers Sustain Salmon-Safe Label. *Capital Press*. August 27.

Hayek, Friedrich A. 1945. The Use of Knowledge in Society. *American Economic Review* 35(4). 519–30.

Heimlich, Ralph E., Jorge Fernandez-Cornejo, William McBride, Cassandra Klotz-Ingram, Sharon Jans, and Nora Brooks. 2000. Genetically Engineered Crops: Has Adoption Reduced Pesticide Use? *Agricultural Outlook*. Washington, D.C.: United States Department of Agriculture. Economic Research Service. August. 13–17. Downloaded from http://www.ers.usda.gov/publications/agoutlook/ aug2000/ao273f.pdf on November 2, 2001.

Heiser, Charles Bixler. 1981. *Seed to Civilization*. Second edition. San Francisco, CA: W.H. Freeman and Company.

Hellmich, R. L., L. C. Lewis, and J. M. Pleasants. 2000a. Survival of Monarch Larvae in Bt and Non-Bt Field Corn. Paper presented at the USDA Monarch Data Review. Chicago, IL. November 16–17.

———. 2000b. Monarch Feeding Behavior and Bt Pollen Exposure Risks to Monarchs in Iowa. Paper presented at the USDA Monarch Workshop. Kansas City, MO. February 24–25.

Hellmich, Richard L., Blair D. Siegfried, Mark K. Sears, Diane E. Stanley-Horn, Michael J. Daniels, Heather R. Mattila, Terrence Spencer, Keith G. Bidne, and Leslie C. Lewis. 2001. Monarch Larvae Sensitivity to *Bacillus thuringiensis*–Purified Proteins and Pollen. *Proceedings of the National Academy of Sciences* 98(21). October 8. 11925–11930.

Henderson, Rick. 1992. Dirty Driving: Donald Stedman and the EPA's Sins of Emission. *Policy Review*. April. 56–60.

Hess, Karl, and Tom Wolf. 1999. Treasure of La Sierra: Colorado's Embattled Taylor Ranch Is the West Writ Small. Here's How Capitalism May Conserve It. *Reason*. October.

Hess, Karl. 2001a. A Tale of Two Earth Days. *TechCentralStation*. April 27.

———. 2001b. Good Intentions Don't Mean Good Conservation. *TechCentralStation*. June 7.

Hill, Peter. 2000. Caution Is the Watchword with Precision Farming. *Farmers Weekly*. June 16. 2.

Hin, C. J. A., P. Schenkelaars, and G. A. Pak. 2001. *Agronomic and Environmental Impacts of the Commercial Cultivation of Glyphosate Tolerant Soybean in the USA*. Utrecht, Netherlands: Center for Agriculture and the Environment. June.

Hulse, Carl. 2002. Effort to Repeal Estate Tax Ends in Senate Defeat. *New York Times Online*. June 12. Downloaded from http://www.nytimes.com/2002/06/13/politics/13ESTA.html on June 13, 2002.

Humane Society of the United States 1999. Choosing a Humane Diet: The Three Rs. Downloaded from http://www.hsus.org/programs/farm/diet/threers.html on November 2, 2001.

Hutcherson, Kate. 1976. Endangered Species: The Law and the Land. *Journal of Forestry*. January. 31–34

IUCN. 2000. *IUCN Red List of Threatened Species*. Gland, Switzerland: IUCN/World Conservation Union.

Jaglo-Ottosen, Kirsten R., Sarah J. Gilmour, Daniel G. Zarka, Oliver Schabenberger, and Michael F. Thomashow. 1998. Arabidopsis CBF1 Overexpression Induces *COR* Genes and Enhances Freezing Tolerance. *Science* 280(5360). April 3. 104–106.

James, Clive. 2001. *Global Status of Commercialized Transgenic Crops: 2000*. ISAA Brief No. 21. Ithaca, NY: International Service for the Acquisition of Agribiotech Applications.

Jehl, Douglas. 2001. Officials Loath to Act as Water Meant for Endangered Fish Flows to Dry Western Farms. *New York Times*. July 9. A8.

Jenner, Mark W. 1998. Coping With the Criminalization of Livestock Production. American Farm Bureau Website. February 19. Downloaded from http://www. fb.com/issues/analysis/livestock.html on July 23, 2001.

Johnson, Nels, Carmen Revenga, and Jaime Echeverria. 2001. Managing Water for People and Nature. *Science* 292(5519). May 11. 1071–1072.

Kades, Deborah. 2001. Your Potatoes Can Be Environmentally Friendly. *Wisconsin State Journal*. April 15. C1.

Kain, Dave. 1996. Botanical Insecticides: Update on Pest Management and Crop Development. *Scaffolds Fruit Journal* 5. July 29. Available at http://www.nysaes. cornell.edu/ent/scaffolds/1996/scaffolds_0729.html.

Kane, Deborah. 2001. The Food Alliance. *Grist Magazine*. May 29. Downloaded from http://www.gristmagazine.com/grist/week/foodalliance052901.stm on June 6, 2001.

Kasuga, Mie, Qiang Liu, Setsuko Miura, Kazuko Yamaguchi-Shinozaki, and Kazuo Shinozaki . 1999. Improving Plant Drought, Salt, and Freezing Tolerance by Gene Transfer of a Single Stress-Inducible Transcription Factor. *Nature Biotechnology* 17(3). March. 287–291.

Kelley, Matt. 2001. Bush Trims Agriculture Budget Cuts Concern Some Midlands Lawmakers. *Omaha World Herald*. April 11.

Kelly, C. David, editor. 1997. *Farm Facts*. Park Ridge, IL: American Farm Bureau.

Kelly, David. 1997. Environment-friendly Conservation Tillage a Growing Practice on America's Farms. American Farm Bureau. October 24. Downloaded from http://www.fb.com/issues/analysis/tillage.html on July 23, 2001.

Kennedy, Robert F., Jr. 1999. I Don't Like Green Eggs and Ham! *Newsweek*. April 26.

Kiewel, Leesa. 2001. Klamath Basin Farmers Defend Their Livelihoods. *AgriNews* 33(48). 1.

Kilman, Scott. 2001. Monsanto Co. Shelves Seed That Turned Out to Be a Dud of a Spud. *Wall Street Journal*. March 21. B21.

King. Tim. 2001. Little Fire on the Prairie. *CityPages.com*. Downloaded from http://www.citypages.com/databank/22/1067/article9552.asp on November 27, 2001.

Kinsella, James. 2001. Wealthy Donors Save Rare Acreage. *Cape Cod Times*. July 27.

Koskinen, W. C., and C. G. McWhorter. 1986. Weed Control in Conservation Tillage. *Journal of Soil and Water Conservation* 41. 365–370.

Laatz, Joan. 1994. Rancher Leases Water Rights to Keep Stream Full of Salmon. *The Oregonian*. June 19.

Landry, Clay. 1998. *Saving Our Streams through Water Markets*. Bozeman, MT: PERC. 6–12.

———. 2002. Unplugging the Everglades. In *Government vs. the Environment*, ed. Donald R. Leal and Roger E. Meiners. Lanham, MD: Rowman & Littlefield.

Leal, Donald R., and J. Bishop Grewell. 1999. *Hunting for Habitat: A Practical Guide to State-Landowner Partnerships*. Bozeman, MT: PERC.

Leal, Donald R., and Roger Meiners, eds. 2002. *Government vs. the Environment*. Lanham, MD: Rowman and Littlefield.

Lee, Mike. 2000. Farmers Turn to "Sustainable" Label for Goods. *Tri-City Herald.* July 29. Downloaded from http://www.tri-cityherald.com/news/2000/0729/story3.html on July 23, 2001.

Lehmann, Volker, and Walter A. Pengue. 2000. Herbicide Tolerant Soybean: Just Another Step in a Technology Treadmill? *Biotechnology and Development Monitor* 43(9). September. 11–14.

Leopold, Aldo. 1991 [1934]. Conservation Economics. In *The River of the Mother of God and Other Essays by Aldo Leopold,* ed. Susan L. Flader and J. Baird Callicott. Madison, WI: University of Wisconsin Press. 193–202.

Lomborg, Bjorn. 2001. *The Skeptical Environmentalist.* Cambridge, UK: Cambridge University Press.

Long, Ed. 2000. Patch Spray Method Might Slash Sugar Beet Herbicide Bills. *Farmers Weekly.* October 20.

López-Bucio, José, Octavio Martínez de la Vega, Arturo Guevara-García, and Luis Herrera-Estrella. 2000. Enhanced Phosphorus Uptake in Transgenic Tobacco Plants that Overproduce Citrate. *Nature Biotechnology* 18(4). April. 450–453.

Losey, John, Linda S. Rayor, and Maureen E. Carter. 1999. Transgenic Pollen Harms Monarch Larvae. *Nature* 399(6733). May 20. 214.

Lueck, Dean, and Jeffrey Michael. 1999. Preemptive Habitat Destruction under the Endangered Species Act. Unpublished manuscript. Bozeman, MT: Department of Agricultural Economics and Economics, Montana State University.

Lueck, Dean. 2000. The Law and Politics of Federal Wildlife Preservation. In *Political Environmentalism*, ed. Terry L. Anderson. Stanford, CA: Hoover Institution Press. 61–119.

MacKerron, D. K. L., J. M. Duncan, J. R. Hillman, G. R. Mackay, D. J. Robinson, D. L. Trudgill, and R. J. Wheatley. 1999. Organic Farming: Science and Belief. In *Scottish Crop Research Institute Annual Report 1998/99.* Dundee, Scotland: Scottish Crop Research Institute. 60–72.

Main, Martin B., Fritz M. Roka, and Reed F. Noss. 1999. Evaluating Costs of Conservation. *Conservation Biology* 13(6). 1262–72.

Malthus, T. R. 1993. *An Essay on the Principle of Population.* Geoffrey Gilbert, ed. Oxford/New York: Oxford University Press.

Mann, Charles C. 1999. Crop Scientists Seek a New Revolution. *Science* 283(5400). January 15. 310–314.

McCallum, Matt, and Greg Brown. 2001. Going Eco: Some Wisconsin Spuds will Carry New Eco Label. *Spudman: Voice of the Potato Industry.* July/August. Downloaded from http://www.spudman.com/pages/issue01vol_06/01_06_going_eco.html on October 13, 2001.

McCarthy, James E., and Mary Tiemann. 2001. MTBE in Gasoline: Clean Air and Drinking Water Issues. CRS Report for Congress. Report 98-290 ENR.

McClelland, Peter D. 1997. *Sowing Modernity: America's First Agricultural Revolution.* Ithaca, NY: Cornell University Press.

McGraw, Linda, and Don Comis. 2000. Standing Crop Residue for Erosion Control. *Agricultural Research.* July. 14–15.

McHughen, Alan. 2000. *Pandora's Picnic Basket: The Potential and Hazards of Genetically Modified Foods*. New York: Oxford University Press.

McNeely, Jeffrey A., and Sara Scherr. 2001 *Common Ground, Common Future: How Ecoagriculture Can Help Feed the World and Save Wild Biodiversity*. Washington, D.C.: IUCN and Future Harvest. May.

Miflin, Ben J. 2000. Crop Biotechnology: Where Now? *Plant Physiology* 123(1). May. 17–27.

Miller, Brian. 1998. Fat of the Land: New York's Waste. *Social Research* 65(1). 75–100.

Miller, Henry I. 2001. Biotechnology, Land Use, and the Environment. In *Agriculture and the Environment*, ed. Terry L. Anderson and Bruce Yandle. Stanford, CA: Hoover Press. 105–131.

Moffat, Anne Simon. 2000. Can Genetically Modified Crops Go "Greener"? *Science* 290(5490). October 13. 253–254.

———. 2001. Finding New Ways to Fight Plant Diseases. *Science* 292(5525). June 22. 2270–2273.

Monarch Watch. 2001. Conservation. Monarch Watch website. Downloaded from http://www.monarchwatch.org/conserve/index.htm on August 14, 2001.

Montana Stockgrower. 2001. Tom and Mary Kay Milesnick of Belgrade Are 'Ranching with Nature.' Spring.

Morriss, Andrew, Bruce Yandle, and Roger Meiners. 2001. The Failure of EPA's Water Quality Control Reforms. *UCLA Journal of Environmental Law & Policy* 20. Forthcoming, December.

Moscatello, Jonathon. 2001. The Food Alliance. *Grist Magazine*. May 30. Downloaded from http://www.gristmagazine.com/grist/week/foodalliance053001.stm on June 6, 2001.

Munkvold, G. P., R. L. Hellmich, and L. G. Rice. 1999. Comparison of Fumonisin Concentrations in Kernels of Transgenic Bt Maize Hybrids and Non-transgenic Hybrids. *Plant Disease* 83(2). February. 130–138.

Murray, Jacqueline. 1970. *The First European Agriculture*. Edinburgh, U.K.: Edinburgh University Press.

National Library of Medicine. 2001. Hazardous Substances Data Bank. Toxicology Data Network website. Downloaded from http://toxnet.nlm.nih.gov/cgi-bin/sis/search/f?./temp/~AAA94ay9P:2 on November 30, 2001.

National Public Radio. 2001. Analysis: Genetically Modified Salt-Tolerant Tomato. National Public Radio program, *Talk of the Nation*. Featuring host Ira Flatow and Professor Eduardo Blumwald of the University of California at Davis. August 3.

National Research Council. 1987. *Introduction of Recombinant DNA-Engineered Organisms in the Environment: Key Issues*. Washington, D.C.: National Academy Press.

———. 1989. *Field Testing Genetically Modified Organisms: Framework for Decisions*. Washington, D.C.: National Academy Press.

Nature Conservancy. 2000. Conservation Beef. November 17. Downloaded from http://nature.org/wherewework/northamerica/states/montana/news/news208.h tml on October 13, 2001.

———. 2001. The Nature Conservancy Saves One of the Largest Unprotected Blocks of Open Space on Martha's Vineyard. Nature Conservancy press release. July 27. Downloaded from http://nature.org/aboutus/press/press322.html on January 28, 2002.

Nixon, Will. 1998. The Color of Money: Cashing In on Green Business. *Amicus Journal* 20(2). Summer. 16–18.

Norton, Seth. 1998. Property Rights, the Environment, and Economic Well-Being. In *Who Owns the Environment?* ed. Peter J. Hill and Roger E. Meiners. Lanham, MD: Rowman and Littlefield. 37–54.

———. 2002. Population Growth, Economic Freedom and the Rule of Law. *PERC Policy Series* PS-24. Bozeman, MT: PERC. February.

O'Malley, Michael. 1997. Dairyman Makes Rounds at Breweries: Revived Beer Industry Provides Waste Grain Fed to Milk Cows. *The Plain Dealer* (Cleveland). November 23.

Oberhauser, Karen S., Michelle D. Prysby, Heather R. Mattila, Diane E. Stanley-Horn, Mark K. Sears, Galen Dively, Eric Olson, John M. Pleasants, Wai-Ki F. Lam, and Richard L. Hellmich. 2001. Temporal and Spatial Overlap between Monarch Larvae and Corn Pollen. *Proceedings of the National Academy of Sciences* 98(21). October 8. 11913–11918.

Oerke E.-C., A. Weber, H.-W. Dehne, and F. Schonbeck. 1994. Conclusion and Perspectives. In *Crop Production and Crop Protection: Estimated Losses in Food and Cash Crops*, ed. E.-C. Oerke, H.-W. Dehne, F. Schonbeck, and A. Weber. Amsterdam: Elsevier. 742–70.

Oregon Water Trust. 1996. Water Right Holder Profile. *Fish Flow News*. Spring.

Organic Trade Association. 2001. Environmental Facts. Organic Trade Association website. Downloaded from http://www.ota.com/environmentalfacts.htm on August 10, 2001.

Organization for Economic Cooperation and Development. 2001. *The Application of Biotechnology to Industrial Sustainability*. Paris: Organization for Economic Cooperation and Development.

Pacific Rivers Council. 2001. *Salmon-Safe Farm Management Certification Program Field Assessor's Guidelines 2.0*. January.

Painter, Kathleen M., and Douglas L. Young. 1994. Environmental and Economic Impacts of Agricultural Policy Reform: An Inter-Regional Comparison. *Journal of Agricultural and Applied Economics*. 26(2). December. 451–462.

Parrott, Wayne. 2001. RR soybean. Posting to AgBioView Listserv. June 4, 2001. Available at http://www.agbioworld.org/listarchive/view.php?id=161.

Pennisi, Elizabeth. 2001. The Push to Pit Genomics against Fungal Pathogens. *Science* 292(5525). June 22. 2273–2274.

PERC. 1998. "Prairie Restorations." *PERC Reports* 16(4):11.

Perks, Bea. 2001. Anti-GM Forces Guilty of "Scientific Apartheid," *BioMedNet News*. July 11. Available at http://news.bmn.com/news/story?day=010712&story=1.

Perrin, Richard, and Terry J. Klopfenstein. 2000. Economic Impact of Feeding West Grain Processors' Byproducts in Nebraska. Lincoln, NE: Institute of Agricultural and Natural Resourcesn University of Nebraska.

Pimentel, David S., and Peter H. Raven. 2000. Bt Corn Pollen Impacts on Nontarget Lepidoptera: Assessment of Effects on Nature. *Proceedings of the National Academy of Sciences* 97(15). July 18. 8198–8199.

Pleasants, J. M., R. L. Hellmich, and L. C. Lewis. 1999. Pollen Deposition on Milkweed Leaves under Natural Conditions. Paper presented at the Monarch Butterfly Research Symposium. Chicago. November 2.

Pleasants, John M., Richard L. Hellmich, Galen P. Dively, Mark K. Sears, Diane E. Stanley-Horn, Heather R. Mattila, John E. Foster, Thomas L. Clark, and Gretchen D. Jones. 2001. Corn Pollen Deposition on Milkweeds in and near Cornfields. *Proceedings of the National Academy of Sciences* 98(21). October 8. 11919–11924.

Postrel, Virginia. 1998. *The Future and Its Enemies.* New York: Simon and Schuster.

Powers, Susan, David Rice, Brendan Dooher, and Pedro Alvarez. 2001. Will Ethanol Blended Gasoline Affect Groundwater Quality? *Environmental Science & Technology.* American Chemical Society. 35(1): 1–48.

Prakash, C. S. 2000. Intellectual Capital: Hungry for Biotech. Technology Review. July/August. Available at http://www.techreview.com/magazine/jul00/prakash.asp.

Pray, Carl, Danmeng Ma, Jikun Huang, and Fangbin Qiao. 2001. Impact of Bt Cotton in China. *World Development* 29(5). May. 813–825.

Progressive Farmer. 2000. Can GM Hybrids Stop Western Corn Rootworms? *Progressive Farmer.* September 11. Available at http://www.biotech-info.net/hybrids_stop.html.

———. 2001. Win-Win Situation: An Innovative Wastewater Plan Boosts Crops and Cleans Up a River. *Progressive Farmer.* January. Downloaded from http://progressivefarmer.com/issue/0101/wastewater/default.asp on January 18, 2001.

Puckett, Bettina. 2001. Localities, Plants Getting the Treatment. *Daily News-Record* (Harrisonburg, VA). August 16.

Raven, Peter, Ray F. Evert, and Susan E. Eichhorn. 1992. *Biology of Plants.* Fifth ed. New York: Worth Publishers.

Rebuffoni, Dean. 1992. Bottom Line Points to Prairie: Corporate Officials See Value in Restoration. *Minneapolis Star Tribune.* December 21. 1B.

———. 1997. In Midwest, the Yard Goes Native. *Christian Science Monitor.* August 15. Downloaded fromhttp://www.csmonitor.com/durable/1997/08/15/us.4.html on October 17, 2001.

Renewing the Countryside. 2001. Urban Artistry. Downloaded from http://www.mncountryside.org/index.cfm?mthd=stry&s_id=155&selectedR_ID=0 on November 7, 2001.

Richard, Cindy Lynn, and Dan Holman. 2000. *Ecological Impact Assessment.* Ames, Iowa: Council for Agricultural Science and Technology, October 12. Available at http://www.cast-science.org/biotechnology/20001011.htm.

Rissler, Jane, and Margaret Mellon. 1996. *The Ecological Risks of Engineered Crops.* Cambridge, MA: MIT Press.

Roberts, Paul. 1999. The Sweet Hereafter: Our Craving for Sugar Starves the Everglades and Fattens Politicians. *Harper's Magazine*. November. 54–68.

Rose, Carol. 1991. Rethinking Environmental Controls: Management Strategies for Common Resources. *Duke Law Journal* 1. 1–38.

Royal Society of London, the U.S. National Academy of Sciences, the Brazilian Academy of Sciences, the Chinese Academy of Sciences, the Indian National Science Academy, the Mexican Academy of Sciences, and the Third World Academy of Sciences. 2000. *Transgenic Plants and World Agriculture*. Washington, D.C.: National Academy Press.

Ryoo, Doohyun, Hojae Shim, Keith Canada, Paola Barbieri, and Thomas K. Wood. 2000. Aerobic Degradation of Tetrachloroethylene by Toluene-o-xylene Monooxygenase of Pseudomonas Stutzeri OX1. *Nature Biotechnology* 18(7). July. 775–778.

Salmon-Safe. 2001. What Is Salmon-Safe? Downloaded from http://www.salmonsafe.org/main.html on July 22, 2001.

Sandman, Lori. 1999. Chesapeake Milk, Eco-labeled by Environmental Quality Initiative: The Environmental Quality Initiative, for a Clean Environment and Profitable Farms. *Environmental Marketing Conference—Online Proceedings*. December 6. Downloaded from http://www.iatp.org/labels/envcommodities/onlineproceedings.html on September 2, 2001.

Saxena, D., and G. Stotzky. 2000. Insecticidal Toxin from *Bacillus thuringiensis* Is Released from Roots of Transgenic *Bt* Corn *in vitro* and *in situ*. *FEMS Microbiology and Ecology* 33(1). July. 35–39.

———. 2001. Bacillus thuringiensis (Bt) Toxin Released from Root Exudates and Biomass of Bt Corn Has no Apparent Effect on Earthworms, Nematodes, Protozoa, Bacteria, and Fungi in Soil. *Soil Biology & Biochemistry* 33(9). July. 1225–1230.

Saxton, Jim, and Mac Thornberry. 1998. The Economics of the Estate Tax: A Joint Economic Committee Study. Washington, D.C.: Joint Economic Committee, United States Congress. December.

Sayler, Tracy. 2000. Progress Made in Using Biotech to Solve FHB in Wheat and Barley. Information Systems for Biotechnology ISB news report. September. 9–10. Available at http://www.isb.vt.edu/news/2000/Sep00.pdf.

Schmidt, Lisa. 2001. Beef with a Mission. *Montana Farmer-Stockman*. June.

Scott, James C. 1976. *The Moral Economy of the Peasant*. New Haven, CT: Yale University Press.

Sears, Mark K., Richard L. Hellmich, Diane E. Stanley-Horn, Karen S. Oberhauser, John M. Pleasants, Heather R. Mattila, Blair D. Siegfried, and Galen P. Dively. 2001. Impact of Bt Corn Pollen on Monarch Butterfly Populations: A Risk Assessment. *Proceedings of the National Academy of Sciences* 98(21). October 8. 11937–11942.

Sen, Amartya. 1999. *Development as Freedom*. New York: Knopf.

Singer, Max. 1999. The Population Surprise. *Atlantic Monthly* 284(2). August. 22–25.

Smirnoff, Nicholas, and John A. Bryant. 1999. DREB Takes the Stress Out of Growing Up. *Nature Biotechnology* 17(3). March. 229–230.

Smith, Ronald, and Roger Leonard. 2001. Farmers and Public Benefit from Insect-Protected Cotton. Alabama Cooperative Extension System *Newsline*. July 9. Available at http://www.aces.edu/dept/extcomm/newspaper/july9a01_op-ed.html.

Soil Association. 2000. *Soil Association Annual Report and Accounts 2000*. Bristol, UK: Soil Association. July 13.

Solley, Wayne B., Robert R. Pierce, and Howard A. Perlman. 1998. Estimated Use of Water in the United States in 1995. U.S. Geological Survey Circular 1200. Denver, CO: U.S. Geological Survey.

Sommerville, Chris, and John Briscoe. 2001. Genetic Engineering and Water. *Science* 292(5525). June 22. 2217.

Squire, G. R., N. Augustin, J. Bown, J. W. Crawford, G. Dunlop, J. Graham, J. R. Hillman, B. Marshall, D. Marshall, G. Ramsay, D. J. Robinson, J. Russell, C. Thompson, and G. Wright. 1999. Gene Flow in the Environment: Genetic Pollution? *Scottish Crop Research Institute Annual Report 1998/99*. Dundee, Scotland: Scottish Crop Research Institute. 45–54.

Staley, Samuel R. 2001. The Political Economy of Land Conversion on the Urban Fringe. In *Agriculture and the Environment*, ed. Terry L. Anderson and Bruce Yandle. Stanford, CA: Hoover Press. 65–80.

Stalling, David. 1999. Public Elk, Private Lands: Should Landowners Benefit from Elk and Elk Hunting? *Bugle*. January–February.

Stanley-Horn, Diane E., Galen P. Dively, Richard L. Hellmich, Heather R. Mattila, Mark K. Sears, Robyn Rose, Laura C. H. Jesse, John E. Losey, John J. Obrycki, and Les Lewis. 2001. Assessing the Impact of Cry1Ab-Expressing Corn Pollen on Monarch Butterfly Larvae in Field Studies. *Proceedings of the National Academy of Sciences* 98(21). October 8. 11931–11936.

Stotzky, G. 2000. Persistence and Biological Activity in Soil of Insecticidal Proteins from *Bacillus thuringiensis* and of Bacterial DNA Bound on Clays and Humic Acids. *Journal of Environmental Quality* 29(3). May–June. 691–705.

Strachan, Alex. 1995. Beer-Swilling Pigs. *Vancouver Sun*. April 26.

Stroup, Richard. 1995. The Endangered Species Act: Making Innocent Species the Enemy. *PERC Policy Series PS-3*. Bozeman, MT: PERC. April.

Sugg, Ike. 1993a. Caught in the Act: Evaluating the Endangered Species Act, Its Effects on Man and Prospects for Reform. *Cumberland Law Review* 24(1). 1–78.

———. 1993b. Ecosystem Babbitt-Babble. *Wall Street Journal*. April 2. A12.

———. 1997. Unsafe for Any Species. *Regulation*. Summer. 12–14.

Takahashi, Michiko, Hiromi Nkanishi, Shinji Kawasaki, Naoko K. Nishizawa, and Satoshi Mori. 2001. Enhanced Tolerance of Rice to Low Iron Availability in Alkaline Soils Using Barley Nicotianamine Aminotransferase Genes. *Nature Biotechnology* 19(5). May. 466–469.

Taylor, O. R. 1999. Monarch Butterflies: Population Dynamics and the Potential Impact of Add-on Mortality. Paper presented at the Monarch Butterfly Research Symposium. Chicago. November 2.

Tilman, David. 1999. Global Environmental Impacts of Agricultural Expansion: The Need for Sustainable and Efficient Practices. *Proceedings of the National Academy of Sciences* 96(11). May 25. 5995–6000.

Times of India. 2001. Salt Tolerant Rice Variety Developed. *Times of India*. September 5. Available at http://timesofindia.indiatimes.com/articleshow.asp?art_id= 272003398.

Trewavas, Anthony. 2001. Urban myths of organic farming. *Nature* 410(6827). March 22. 409–410.

Trewavas, Anthony J., and Christopher J. Leaver. 2001. Is Opposition to GM Crops Science or Politics? *EMBO Reports* 21(6). June. 455–459.

Trice, Calvin R. 2000. Waste Plan a Triumph for All: Nutrient Diverted from Bay to Farms. *Richmond Times-Dispatch*. September 14.

Truini, Joe. 2001. "Serving Up Another Round: Brewers Partner with Farmers, Others to Achieve Zero Waste." *Waste News*. Crain Communication.

Tu, Jumin, Guoan Zhang, Karabi Datta, Caiguo Xu, Yuqing He, Quifa Zhang, Gurdev Singh Khush, and Swapan Kumar Datta. 2000. Field Performance of Transgenic Elite Commercial Hybrid Rice Expressing Bacillus thuringiensis d–endotoxin. *Nature Biotechnology* 18(10). October. 1101–1104.

United Nations Development Program. 2001. *Human Development Report 2001: Making New Technologies Work for Human Development*. New York: Oxford University Press.

United Nations Population Division. 1999. *The World at Six Billion*. Population Division of the Department of Economic and Social Affairs Working Paper ESA/P/WP 154. New York: Population Division of the Department of Economic and Social Affairs of the United Nations Secretariat.

———. 2001. *World Population Prospects: The 2000 Revision. Key Findings*. Downloaded from http://www.un.org/esa/population/publications/wpp2000/ wpp2000h.pdf on October 19, 2001.

United States Department of Agriculture. 1997. *1997 Census of Agriculture*. Washington, D.C.: National Agricultural Statistics Service.

———. 2000a. *Acreage*. Washington, D.C.: U.S. Department of Agriculture, National Agricultural Statistics Service. June 30.

———. 2000b. Production Practices for Major Crops in the U.S. Agriculture, 1990–1997. Economic Research Service Statistical Bulletin, Number 969. 67.

———. 2000c. USDA Forest Service Strategic Plan (2000 Revision.) Washington, D.C.. Downloaded from http://www2.srs.fs.fed.us/strategicplan/sp2000.pdf on June 12, 2001.

———. 2001. *Acreage*. Washington, D.C.: U.S. Department of Agriculture, National Agricultural Statistics Service. June 29.

United States Department of the Interior, Fish and Wildlife Service and U.S. Department of Commerce, Bureau of the Census. 1997. *1996 National Survey of Fishing, Hunting, and Wildlife-Associated Recreation*. FHW/96-NAT. Washington, D.C.. November.

United States Environmental Protection Agency. 1996. Review of the National Ambient Air Quality Standards for Particulate Matter: Policy Assessment of Scientific and Technical Information. OAQPS Staff Paper, US Environmental Protection Agency, EPA-452, R-96-013, July. Downloaded whole report from http:// www.epa.gov/ttn/oarpg/t1sp.html and cited portion of report from http:// www.epa.gov/ttn/oarpg/t1/reports/pmspchs.pdf on December 10, 2001.

————. 1999. Evaluation of the Carcinogenic Potential of Pyrethrins: Final Report of the Cancer Assessment Review Committee. Washington, D.C.: Environmental Protection Agency, Office of Pesticide Programs Health Effects Division. April 8.

————. 2000a. Environmental Assessment. *Bt Plant-Pesticides Biopesticides Registration Action Document*. Environmental Protection Agency Scientific Advisory Panel Preliminary Report. October. Available at www.epa.gov/scipoly/sap/2000/october/brad3_enviroassessment.pdf.

————. 2000b. *Proposed Regulations to Address Water Pollution from Concentrated Animal Feeding Operations*. EPA 833-F-00-016. Washington, D.C.: Office of Water. December.

————. 2001a. *Biopesticides Registration Action Document: Bacillus thuringiensis (Bt) Plant-Incorporated Protectants*. U.S. Environmental Protection Agency Office of Pesticide Programs Biopesticides and Pollution Prevention Division. October 15. Available at http://www.epa.gov/pesticides/biopesticides/reds/brad_bt_pip2.htm.

————. 2001b. "Waste Wi$e: A Sampling of Waste Reduction Results." Downloaded from http://es.epa.gov/partners/wise/wwsamp.html on September 14, 2001.

United States Fish and Wildlife Service. 2001. Summary of Listed Species. Downloaded from http://ecos.fws.gov/tess/html/boxscore.html on July 18, 2001.

United States General Accounting Office. 1994. *Endangered Species Act: Information on Species Protection on Nonfederal Lands*. GAO/RCED-95-16. Washington, D.C.: U.S. General Accounting Office. December.

————. 1997. *Tax Policy: Effects of the Alcohol Fuels Tax Incentives*. GAO/GGD-97-41. Washington, D.C.: U.S. General Accounting Office. March.

————. 2000. *Sugar Program: Supporting Sugar Prices Has Increased Users' Costs While Benefiting Producers*. GAO/RCED-00-126. Washington, D.C.: U.S. General Accounting Office. June.

————. 2001. *Farm Programs: Information on Recipients of Federal Payments*. GAO-01-606. Washington, D.C.: U.S. General Accounting Office. June.

United States House. 2000. Death Tax Elimination Act of 2000. *Congressional Record*, 106th Cong., 2nd sess., H4158. June 9.

Vasey, Daniel E. 1992. *An Ecological History of Agriculture*. Ames: Iowa State University Press.

Vitousek, Peter M., Harold A. Mooney, Jane Lubchenco, and Jerry M. Melillo. 1997. Human Domination of Earth's Ecosystems. *Science* 277(5325). July 25. 494–499.

Vörösmarty, Charles J., Pamela Green, Joseph Salisbury, and Richard B. Lammers. 2000. Global Water Resources: Vulnerability from Climate Change and Population Growth. *Science* 289(5477). July 14. 284–288.

Weed, Becky. 2000. What Are the Nonlethal Control Methods and How Effective Are They? Thirteen Mile Lamb and Wool Company Website. Downloaded from http://www.lambandwool.com/nlethalc.html on October 13, 2001.

Welch, Craig. 2001. Both Sides Harden in Oregon Water Dispute. *Seattle Times*. July 9. A1.

Westwood, Jim. 1999. Virus-Resistant Papaya in Hawaii: A Success Story. Information Systems for Biotechnology *ISB News Report*. January. 6–7. Available at http://www.isb.vt.edu/news/1999/Jan99.pdf.

White, K. D. 1970. *Roman Farming*. Ithaca, NY: Cornell University Press.

Wilcove, David S., Michael J. Bean, Robert Bonnie, and Margaret McMillan. 1996. Rebuilding the Ark: Toward a More Effective Endangered Species Act for Private Land. Environmental Defense. December 5. Downloaded from http://www.environmentaldefense.org/pubs/Reports/help-esa/ on February 23, 2001.

Wilkinson, Todd. 1997. More Ranchers Are Declaring a Cease-Fire on the Coyote. *Christian Science Monitor*. September 22.

———. 2000a. Eating as Advocacy. *Orion Afield*. Summer.

———. 2000b. How Fly-fishing Can Save the West's Ranchers. *Christian Science Monitor*. September 7. 1.

Wong, Eric A. 2001. Environmentally Friendly Transgenics. Information Systems for Biotechnology *ISB News Report*. June. 8. Available at http://www.isb.vt.edu/news/2001/Jun01.pdf.

Woodbury, Richard. 2000. The Best Coyote Defense Since the Road Runner. *Time*. February 21.

World Bank. 1997. Bioengineering of Crops Could Help Feed the World: Crop Increases of 10–25 Percent Possible. World Bank Press Release. October 9. Available at http://www.worldbank.org/html/cgiar/press/biopress.html.

World Health Organization. 1991. *Strategies for Assessing the Safety of Foods Produced by Biotechnology: Report of a Joint FAO/WHO Consultation*. Geneva, Switzerland: World Health Organization.

World Wildlife Fund, Wisconsin Potato and Vegetable Growers Association, and University of Wisconsin Extension Service. 2001. Wisconsin Potato Growers to Market under World Wildlife Fund Logo. Press release. May. Downloaded from http://ipcm.wisc.edu/bioipm/news/press_releases/May2001.html on October 13, 2001.

Wornson, George. 1989. Pollution Source Reduction in Food Processing: Looking for a Beneficial Use for Everything. Miller Brewing Company. August 22.

Zangerl, A. R., D. McKenna, C. L. Wraight, M. Carroll, P. Ficarello, R. Warner, and M. R. Berenbaum. 2001. Effects of Exposure to Event 176 *Bacillus thuringiensis* Corn Pollen on Monarch and Black Swallowtail Caterpillars under Field Conditions. *Proceedings of the National Academy of Sciences* 98(21). October 8. 11908–11912.

Zhang, Hong-Xia, and Eduardo Blumwald. 2001. Transgenic Salt-Tolerant Tomato Plants Accumulate Salt in Foliage but Not in Fruit. *Nature Biotechnology* 19(8). August. 765–768.

Index